The Making and Evaluation
of Holograms

The Making and Evaluation of Holograms

NILS ABRAMSON

Department of Production Engineering,
Royal Institute of Technology, S-100 44 Stockholm, Sweden

1981

ACADEMIC PRESS

A Subsidiary of Harcourt Brace Jovanovich, Publishers

London • New York • Toronto • Sydney • San Francisco

ACADEMIC PRESS INC. (LONDON) LTD
24-28 Oval Road,
London NW1

United States Edition published by
ACADEMIC PRESS INC.
111 Fifth Avenue
New York, New York 10003

British Library Cataloguing in Publication Data
Abramson, N.
   The making and evaluation of holograms.
   1. Holography
   I. Title
   774   QC449

   ISBN 0-12-042820-2
   LCCCN 81 0521

Typeset by Advanced Filmsetters (Glasgow) Ltd
and printed in Great Britain by Thomson Litho Ltd, East Kilbride, Scotland

Preface

The physical process of holography was first realized by Gabor in 1947.[1] However, the quality of his images was inferior to that of ordinary photography and as a result he did not succeed in arousing much interest in his work. The main reason for the discouraging quality of his images was that no light source existed with the combined intensity and coherence that was needed.

In 1917 Einstein had predicted the existence of the stimulated emission of radiation which subsequently made possible the invention of the laser.[2] This new type of radiation was proposed simply to satisfy Einstein's equations: in order to establish equilibrium between the absorption and the emission of spectral lines, he had to introduce a stimulated emission which would amplify the radiation of certain spectral lines. Not until 1960, long after Einstein's death, was his discovery realized in practice with the development of the first laser (Light Amplification by Stimulated Emission of Radiation).

When laser light became available, Leith and Upatnieks applied it to holography, which they had reinvented and developed further.[3] Their experiments resulted in excellent three-dimensional images that astonished the scientific community. Gabor, who by now was working in utterly different fields, was awarded the Nobel prize in 1971. Thus, the combination of Einstein's prediction of stimulated emission and Gabor's prediction of holography resulted in the sophisticated imaging and measuring methods which will be described in this book.

Among others, Powell and Stetson[4] in America and Burch and his collaborators[5] in England discovered interference fringes in the holographic images of objects that were moved during exposure of the hologram. Very soon several new methods were developed[6,7] that utilized such fringes for measurement. These techniques came to be known as holographic interferometry.

This book will describe how holographic interferometry can be used for the measurement of dimensions, deformations and vibrations. By counting the interference fringes and by studying how the fringe pattern changes when the angle of observation is altered, it is possible to evaluate both the direction and the amplitude of a displacement. To explain the process of fringe formation a number of analogies will be used, such as the holo-diagram, moiré pattern, string, mirror, and grating analogies, the analogy of reflections from motion trails, and finally the analogy between light of short coherence length and of short pulse duration.

Although mathematics is perhaps the most general method of analogy, and can be used within a very broad spectrum of fields, because of its very generality it sometimes becomes more complicated than the phenomenon which it tries to explain. It is the author's opinion that in such cases other and suitable analogies should be used, and thus in this book mathematics has been utilized mainly to verify and to prove the correctness of alternative methods of analogy. A further reason for not using mathematics to explain the behaviour of interference patterns is that these patterns represent direct visual information: the best form of explanation is therefore a comparison with analogous visual information that is easier to understand intuitively, as for example moiré patterns that are simply built up by intersecting lines.

Suppose for a moment that holography produced not patterns but rather columns of digits bearing the corresponding information capacity. Surely people would in that case ask themselves what could be done with all these numbers? The goal would then necessarily be to design a graphical display which would transform the numbers into a type of map, the contour lines of which would represent, for example, lines of constant displacement or constant vibration amplitude. This is exactly what hologram interferometry can do directly, all by itself.

Chapter 1 introduces the reader to the concept of the coherence of light, the formation of interference fringes, and what happens to the light energy when two light beams extinguish each other resulting in darkness. It is shown that the addition of coherent light intensities is a non-linear process and that therefore an interferometer can be used in a limited way as a light intensity amplifier. A diagram explains this amplification of a signal beam as a function of a reference beam. The accepted formula for the "diffraction-limited" and the "interference-limited" resolution of optical systems is derived, based on the simple concept of energy conservation. The analogy between interference fringes and moiré fringes is described, and the discussion in the rest of the book will to a large degree depend on an extended use of this analogy. Some instruments based on the principles of interferometry are described and their functions explained. One of these is the "interferoscope", a recently developed instrument with novel possibilities for fringe manipulation. The reader is introduced to the holo-diagram and to the new concept "wavefronts of observation".

In *Chapter 2* the holographic process is explained and the true and false holographic images are studied. A new equation and a new graphical method are presented which predict whether an image is to be virtual or real. It is shown that both the virtual and the real images fall on to a straight line through the point source of the reference beam. Holography is compared to photography, and it is shown why the former can produce a three-dimensional image containing a higher amount of information. Holographic memories are discussed, and it is demonstrated that for a given size of the components their information contents will not exceed those of photographic memories. Speckles are discussed as being optical information quanta, the number of which determine the information content. Finally, different holographic methods and configurations are examined.

The use of holography for the measurement of dimensions, deformations and vibrations is studied in *Chapter 3*. The most obvious way of making measurements in the holographic image is to look through the hologram plate as if it was a "window with a memory", and to use conventional optical instruments for determining the positions of object points. A new moiré pattern diagram is demonstrated which, in a simple way, visualizes the optical resolution in the space around an observing lens. However, holographic methods exist the precision of which is not restricted by the diffraction-limited resolution of observing instruments but only by the interference-limited resolution of the hologram plate. These methods are all the result of interferometric comparison of an object at different situations. The methods described are real-time hologram interferometry, double-exposure holography, time-averaged holography and sandwich holography. The last method is treated in some detail because it is not as well known as the others. Different holographic configurations are described, and the stability requirements of different components are studied. The advantage of simplicity in the holographic set-up is emphasized.

Chapter 4 describes the evaluation of fringes caused by out-of-plane motion. The fringe-forming process is discussed and the importance of defined path lengths is emphasized. It is shown that the light should pass just one scattering surface when hologram interferometry is used in a general way. Trigonometric methods for fringe evaluation are described, and it is shown that a cone of constant sensitivity exists in the direction of motion of the object. It is proved mathematically that maximal sensitivity is represented by the plane which bisects the directions of illumination and observation. Examples are given of how to separate out-of-plane motions from in-plane motions. The concept of Young's fringes of observation is introduced. The moiré pattern analogy to hologram interferometry is explained by substituting the scattering surface of the object by an equivalent grating. A string analogy is studied which is simply based on substituting for the light rays by using stretched strings. Finally, the holo-diagram is re-introduced, and it is shown how it can produce information about the sensitivity distribution in the measuring space.

Chapter 5 explains why ordinary holograms interferometry is not so well suited for the measurement of in-plane motions. The advantages and disadvantages of oblique illumination and observation are described, and the use of two illuminating beams explained. When an ordinary hologram is studied the fringes move as the angle of observation is varied, and from this motion the in-plane displacement can be calculated. This motion sometimes produces a parallax effect, as if the fringe system existed in front of, or behind, the surface of the object. This effect, the "localization" of the fringes, can be used for evaluation and is usually explained as being caused by "homologous" rays. A new explanation of the localization is presented which is based on the focusing effect caused by reflections from "motion trails" and some practical examples are given. The holo-diagram is used to explain complicated fringe patterns. The moiré pattern analogy to hologram interferometry is proved mathematically for every type of motion. A large number of object motions are studied and the resulting interference patterns are visualized by use of the moiré pattern analogy. A bold new statement that hologram interferometry functions without interferometry and even without the hologram is first proposed and then proved. In the case of hologram-free interferometry a lens has to be used to produce an image on which the fringe patterns are distributed. This method is well known as speckle photography. Finally, the in-plane sensitivity of sandwich holography is studied and an equation derived.

Chapter 6 starts with a discussion of the "rose of error", a figure which demonstrates that even if there exists only one fringe pattern for a given deformation there may exist many different deformations which produce the same fringe pattern. It is shown that the single true deformation can be singled out by fringe manipulation using either interference or a moiré pattern effect at the image plane. The former method is much more general and produces more information. It is most easily accomplished by the use of sandwich holography, which can reveal whether the direction is forwards or backwards and can also compensate for unwanted motions of more than 1 mm. Additionally, it can be used to calculate stresses and strains caused by bending without having to measure the derivative of the fringe spacing. Practical examples are presented in the form of milling machines nearly 2 m high and weighing around 2000 kg on which measurements have been carried out which are accurate to a fraction of a thousandth of a millimetre. A static load represents the cutting force, and the resulting deformations are studied even on machine parts that have been displaced by several tenths of a millimetre. A small hand-held drilling machine is studied by double-pulsed holography while in operation and, using a fast spinning sandwich hologram, the large and unwanted motions of the hand-held machine can be compensated for and the local vibrations examined. Finally, some fringe counting rules are presented.

Chapter 7 describes different equipment and methods for holography, starting with the helium–neon, argon and ruby lasers. The coherence of laser light is studied with particular regard to those details that are of importance to holography. The influence of polarization is discussed, and it is shown that it can be used to eliminate some disturbances at the hologram plate. The beam expander is described and calculations of the pinhole size of the spatial filter are given. Despite the fact that the light from a single He–Ne laser has a coherence length limited to about 0.3 m, it is demonstrated that objects many metres long can be holographed. Furthermore, it is shown mathematically and diagrammatically that the coherent area repeats itself with a path length of twice the laser length. This repetivity of the coherence length can be used to produce surfaces of constant distance which intersect the object and produce fringes of constant height on its surface. These fringes are referred to as "contouring fringes" and a number of methods for their production are discussed. It is emphasized that they can appear accidently and be misinterpreted as representing displacement or deformation. The short coherence length can be used to advantage by its employment in the method of "light-in-flight recording by holography". It is shown that, in holography, light of short coherence length produces a result that is identical to that of a short light pulse. This analogy between coherence and pulse length makes possible a whole new field of holographic visualization, in the form of a "movie" picture, of light pulses propagating through optical components. Using short light pulses it is possible also to study other ultra-high-speed phenomena. Finally, at the end of the book practical advice is given as to the stability requirements of the holographic set-up and its components. An example is cited of an object 1 m long which, after a first exposure had been made of it using a sandwich hologram, was transported by car to a workshop for machining. Some days later it was brought back to the holographic set-up and a second exposure was made, producing excellent interference fringes revealing the deformations in the object. This experiment is just one of many described in the book that demonstrate that holography is now well suited for making practical measurements in workshop conditions.

Stockholm   N.A.
February 1981

References

1. D. Gabor. Microscopy by reconstructed wavefronts. *Proc. R. Soc. A* **197**, 454 (1949).
2. A. Einstein. Zur Quantentheorie der Strahlung. *Phys. Z.* **18**, 121 (1917).

3. E. Leith and J. Upatnieks. Reconstructed wavefronts and communication theory. *J. opt. Soc. Am.* **52**, 1123 (1962).
4. R. Powell and K. Stetson. Interferometric analysis by wavefront reconstruction. *J. opt. Soc. Am.* **55**, 1593 (1965).
5. J. Burch. The application of lasers in production engineering. *Prod. Engng* **44**, 431 (1965).
6. R. Collier, E. Doherty and K. Pennington. Application of moiré techniques to holography. *Appl. Phys. Lett.* **7**, 223 (1965).
7. P. Brooks, L. Heflinger and R. Weurker. Interferometry with a holographically reconstructed comparison beam. *Appl. Phys. Lett.* **7**, 248 (1965).

Contents

1

Interference

1.1 The formation of interference fringes

If an ordinary incandescent lamp is switched on, the room becomes bright. If another lamp is switched on, the room becomes brighter. If the two lamps are of equal intensity and are placed at equal distances from a screen, this screen will receive about an equal amount of energy from each lamp. Every point will receive about double the light energy received if only one lamp is used (Fig. 1.1(a)). It would be surprising if this were not the case.

On the other hand, if a semi-transparent mirror divides the light from a laser into two beams which thereafter coincide on a screen, some points on that screen may be darker when the two beams illuminate the screen simultaneously than if just one beam is used. The two beams may even result in darkness, as seen in Fig. 1.1(b), or the whole beam may disappear completely if an interferometer is used, as in Fig. 1.1(c). In Fig. 1.1(c) a laser at the left emits a beam, one part of which (I) passes straight through the two semi-transparent mirrors a and d, while another part (II) is reflected by the mirrors a, b and c, whereafter it is reunited with the first part after a fourth reflection by d. It can be arranged that if beam I is stopped, e.g. by putting a hand between a and b, there will be a certain intensity at the screen, but if beam I is permitted to pass freely between a and d the intensity at the screen becomes zero. What has happened to the light energy? This will be dealt with later in this book.

Another curious phenomenon of laser light can be studied without any instruments at all. If a diffuse surface is illuminated by a laser, the illumination does not appear to be uniform but rather consists of randomly spaced bright and dark spots or speckles.

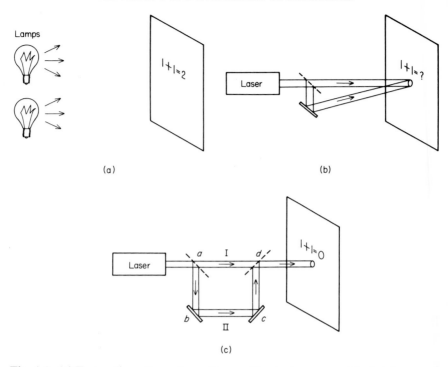

Fig. 1.1. (a) Two ordinary incandescent lamps illuminate a screen. The brightness on this screen is higher when both the two lamps are switched on than if only one is working. It would be very surprising if it had been otherwise. (b) A laser beam is divided into two by a beam-splitter and both beams fall on the screen. At some points on the screen the brightness will be lower when the two beams illuminate simultaneously than if only one of the beams is allowed to reach the screen. What has happened to the energy? (c) The beam from a laser is split in two parts (I and II) which travel by different paths before they are recombined. The strange situation can occur that if one beam is let through at a time there will be brightness on the screen, but when the two beams illuminate the screen simultaneously their intensities add up to darkness! Has the energy disappeared? *a* and *d* are beam-splitters (semi-transparent mirrors) whereas *b* and *c* are ordinary mirrors.

1.1.1 Introduction to the coherence of laser light

These strange results are all caused by the fact that all the waves in a cross-section through the laser beam are oscillating in phase (the light is transversely coherent), that all the waves have equal wavelength (the light is mono-chromatic) and that the light source, if uninterrupted, emits a continuous form of waves (the light is time-coherent). The result of these ideal qualities (which in practice are never completely fulfilled) is that the laser light can be regarded as a perfect sinusoidal wave.

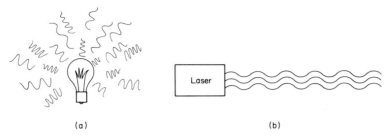

(a) (b)

Fig. 1.2. (a) The different waves from an ordinary incandescent lamp are radiating in a random way from the individual atoms, which are independent of each other. Thus there is no connection between their wavelengths or their phases. (b) The light waves from a laser, on the other hand, are produced by a well organized oscillator, and thus all the waves are in phase with one another and have a single wavelength: the light is coherent.

Let us compare the light from an ordinary bulb with that from a laser, as shown in Fig. 1.2(a) and (b). The incandescent lamp produces its light simply by heat, i.e. the atoms of its tungsten filament are shaking around so violently that the electrons change their orbits at random around the nucleus. Each time that an electron falls to a lower (less energetic) orbit an atom loses energy and emits a damped wave of electromagnetic oscillation representing an extremely short flash of light. Thus there is no relation in direction, wavelength or phase between the waves of the different flashes.

The laser, on the other hand, emits smooth uninterrupted waves that all

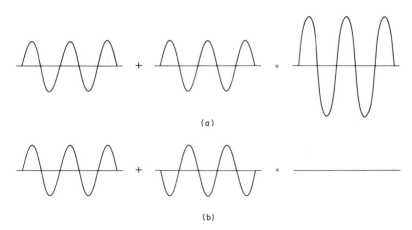

(a)

(b)

Fig. 1.3. The light waves from a laser behave like radio waves or any a.c. voltage. The electrical component oscillates in a sinusoidal way between positive and negative. When two waves are in phase their amplitudes are added—constructive interference (a); when the two waves are out of phase their amplitudes are subtracted—destructive interference (b).

have the same direction and wavelength and oscillate in phase. If light could be heard by the human ear, that from an incandescent lamp might sound like the traffic noise on a busy street, while that from a laser would sound like a single note on a violin.

Let us now examine more closely the wave nature of light, which is a transverse electromagnetic oscillation; this means that it is vibrating in a plane perpendicular to the direction of travel. The fact that light can be polarized proves this fact. Two laser beams can cancel each other only if they have identical planes of polarization.

Figure 1.3 demonstrates what happens when two identically polarized coherent laser beams of the same intensity coincide. If the two beams are in phase their amplitudes (measured, for example, in volts) are added; if they are totally out of phase the sum of their voltage amplitudes will be zero, so there will be no resulting light intensity at all. The amplitudes simply behave as do those of any a.c. voltages that are added to each other, independently of the fact that light has a frequency of the order of 10^{14} Hz whereas the frequency of ordinary household electrical current is 50–60 Hz.

1.1.2 Analogies

Initially let us study only the two cases in which the waves are either totally in phase or totally out of phase. Thus the situation is simplified, and the sinusoidal oscillation of light (Fig. 1.4(a)) can be approximated into a square wave oscillation (Fig. 1.4(b)). Later on the light waves will be represented by dark and white stripes (Fig. 1.4(c)), where the white areas represent positive voltage and the shaded areas negative voltage.

The interaction of two light beams will be represented by the moiré pattern resulting from placing one such pattern on top of the other. The *constructive* interference of Fig. 1.3(a) then corresponds simply to the effect of super-imposing two patterns of the type shown in Fig. 1.4(c) such that the white parts of each pattern correlate exactly. The *destructive* interference of Fig. 1.3(b) then corresponds to displacing a white area sideways such that the dark stripes of one pattern cover the bright stripes of the other. In that case the whole beam will appear to be of the same level of darkness all the way along, and no sign of the waves will be evident.

Admittedly there is one drawback to this analogy: the pattern resulting from two identical sets of stripes placed exactly on top of each other produces an unchanged pattern which should represent constructive interference; however, there is no component in the moiré pattern that corresponds to the sum of the brightnesses of the two real light beams. Fortunately this dis-advantage can be accepted because we are not interested in the intensities of the interference fringes but only in their position and separation.

The description of interference phenomena given here has borrowed many

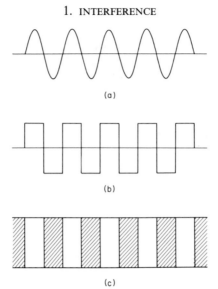

Fig. 1.4. The moiré pattern analogy to interferometry is based on a situation in which the sinusoidal wave (a) is approximated into a square wave (b), which is represented by alternatively bright and dark fringes (c). One wavelength then consists of one bright and one dark fringe.

terms from electronics. Thus a fringe pattern could be said to represent a frequency like that of a.c. current, but in the spatial domain instead of in the time domain. A close fringe spacing represents a high spatial frequency, broad fringes represent a low frequency, and finally no fringes at all represent spatially d.c. light.

In electronics any complex waveform can be split up into a number of sine waves of different frequencies (Fourier transformation). The corresponding effect in optics is even simpler as the Fourier transformation occurs spontaneously. In optics an electronic step function would be represented by a sharp edge intersecting a laser beam. At short distances there will be a sharp shadow but at long distances there will be a fringe pattern representing its Fourier transform.

1.1.3 Two intersecting laser beams

Interference fringes are formed if a laser beam is split into two beams which are directed in such a way that they intersect at an angle. The moiré pattern analogy to this phenomenon is demonstrated in Fig. 1.5. The two light beams are directed from A to D and from B to C, respectively. The straight wavefronts are represented by equally spaced straight lines perpendicular to the direction of the light. One dark plus one bright line represents one wave-

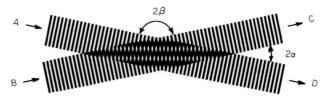

Fig. 1.5. Two coherent light beams, directed from A to D and from B to C respectively, intersect at an angle 2α. Interference fringes are formed that bisect this angle. The directions and angles of the interference fringes are analogous to the moiré fringes shown in the figure.

length, and the moiré pattern at the point of intersection is analogous to the interference pattern. This is true both for the separation and for the direction of the fringes.

1.1.4 Path length differences

Let us now take a closer look at the fringe pattern (Fig. 1.5). If a point moves along one interference fringe from left to right, the distance to A and the distance to B will increase by the same amount because the interference fringe bisects the angle between AD and BC. (When the point has passed one diagonal of an interference rhomb the path length of each beam has increased by 0.5λ, where λ is the wavelength of the light.) Thus the phase difference at the intersection of the two beams is constant along one interference fringe.

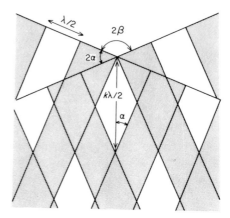

Fig. 1.6. When the two beams of Fig. 1.5 intersect, rhombs are formed whose two diagonals have the lengths $0.5\lambda/\sin\alpha$ and $0.5\lambda/\cos\alpha$ respectively. These two values are referred to in this book as $k^*\lambda/2$ and $k\lambda/2$ and represent not only the fringe spacing but also the diffraction-limited and the interference-limited resolutions of optical systems.

If a point moves from one interference fringe to an adjacent fringe, one beam is shortened by 0.5λ while the other is lengthened by the same amount. In this case, the path length changes by one wavelength (λ) and the phase difference at the object changes accordingly by 2π radians (360°).

If a point moves along one interference fringe from left to right (Fig. 1.5), the distance to D will decrease by the same amount as the distance to B increases. Thus, the path length of light from B that is reflected by the point in the direction from A to D is not influenced by its motion along the direction of one interference fringe (the bisector of the angle between the beams AD and BC). This is *independent of whether the beam AD exists or not.*

If the point moves upward from one interference fringe to an adjacent fringe, the distance to B and the distance to D will both increase by 0.5λ. In that case, the path length of the light from B that is reflected by the point in the direction from A to D changes by one wavelength (λ) and the phase at D accordingly changes by 2π radians (360°). This result is also independent of whether the beam AD exists or not.

The dark parallel fringes are separated by one diagonal ($k\lambda/2$) of the bright rhombs. From Fig. 1.6 it can be shown that

$$k^* = \frac{1}{\sin \alpha} \tag{1.1}$$

or, if we use the angle β instead,

$$k = \frac{1}{\cos \beta}, \tag{1.2}$$

where $k\lambda/2$ is the distance between adjacent fringes, λ is the wavelength of light, α is half the angle between the two beams, $\beta = 90 - \alpha$.

The opposite diagonal of the rhombs represents the direction of the fringes which bisect the angle 2α because of the symmetric properties of the rhombs. If the two beams move at the same speed from A to D and from B to C, respectively, the fringe system will remain stationary. Each rhomb will move to the right, invariably along the bisector of the angle between the beams AD and BC. Therefore, if the wavelength of the two laser beams is constant and equal, and if the speed of the two beams is equal, each fringe will remain stationary. This is a very important fact. Despite the high speed of light ($c.\ 3 \times 10^8\,\mathrm{m\,s^{-1}}$) and its high frequency ($c.\ 10^{14}\,\mathrm{Hz}$), slow or even static detectors can be used to measure relative changes of phase.

Figure 1.5 is of course only a two-dimensional representation in which the interference fringes exist in the form of equally spaced, flat, parallel bright and dark sheets perpendicular to the surface of the paper. From Fig. 1.5 the same conclusions can be expressed in four different ways:

(1) If two laser beams are directed away from A and B, respectively, the interference fringes will bisect the angle between the beams AD and BC.

(2) If two laser beams are directed towards C and D, respectively, the interference fringes will bisect the angle between CB and DA.

(3) If one laser beam is directed away from A and another is directed towards C the interference fringes will be parallel to the bisector of the angle between AD and BC.

(4) If one laser beam is directed away from B and another is directed towards D, the interference fringes will be parallel to the bisector of the angle between BC and AD.

A plane mirror placed parallel to the fringes would reflect light from A towards C just as if it had come from B. Indeed, if each fringe were replaced by a plane mirror, and if the beam from B were switched off, all the reflections of the beam from A by all mirrors would arrive in phase. The path length of the reflections from one mirror differs from those of adjacent mirrors by one wavelength. The mirrors fulfil the Bragg conditions and equations (1.1) and (1.2) represent the relations between the mirror distance and the Bragg angle.

Let us switch on B again and see what will happen if a semi-transparent mirror is placed in the fringe system of Fig. 1.5. Experiments show that if it is placed with all of its surface within one dark interference fringe, nothing changes at C or D. However, if the mirror is kept parallel to the fringes but moved out into the brightness of one fringe, all the light will be directed either towards C or towards D. If the mirror is placed at an angle to the fringes so that it intersects, for example, five fringes, five fringes will also appear both at C and at D. Dark fringes at C correspond to bright fringes at D because the energy not reaching C will reach D instead. Thus, there exist at least two methods for predicting the fringes at C and D. The conventional way is to calculate the angle between the light passing through the mirror and the light reflected by the mirror and then to find the fringe separation using equation (1.1). Another way is to use (1.1) first and then count the number of inter-sections of the mirror surface and the interference fringes. The advantage of using the latter method will be demonstrated later in this chapter. This is especially important when the semi-transparent mirror is replaced in practice by an object on which measurements are to be made.

Consider now what happens if a photographic plate is placed in the fringe pattern of Fig. 1.5. After exposure, development and fixation, the fringe pattern will be recorded on the plate in the form of bright and dark lines. If the plate is repositioned and illuminated by only the beam from A, the plate will deflect light towards C as if it came from B. The deflection of the light by the grid is known as diffraction, and the grid is called a diffraction grating. The method of producing a grating using a photographic plate illuminated by two or more coherent light beams represents the basic principle of holography. When the photographic emulsion is thick the hologram is a thick hologram, or a Bragg hologram, which "remembers" not only the fringe spacing but also the angles of the intersecting interference surfaces.

The fringes studied until now have been formed by two intersecting beams of laser light in which each wavefront is a flat surface. These beams could be made up of the parallel rays directly from a laser, or they could arrive from point sources at infinite distances, or from other sources the light of which has been made parallel (collimated) by means of lenses or mirrors. A necessary condition for the fringes to be stationary is that the two sources are mutually coherent (have a fixed phase relation). The easiest way of reaching this condition is by letting the two beams emanate from the same light source, as seen in Fig. 1.1(b).

1.1.5 The moiré pattern analogy

Summing up, the analogy between the moiré fringes of Fig. 1.5 and the interference fringes formed by two intersecting beams travelling from A to D and from B to C respectively may be described as follows:

Moiré	*Interference*
Fringes bisect the angle 2α.	Fringes bisect the angle 2α.
The straightness of the fringes depends on the straightness and equal spacing of the stripes.	The straightness of the fringes depends on the straightness and constant spacing of the wavefronts.
If the two striped bands are moving with identical velocities from A to D and from B to C, the moiré patterns stay fixed.	As the wavefronts move with the speed of light (c), the fringes stay fixed.
If the striped band AD is moved towards D one more stripe than BC is moved towards C, then the fringes will move downwards one fringe spacing.	If the phase of the beam AD is advanced one wavelength (2π rad) compared to BC, then the fringes will move downwards one fringe spacing.
The fringe spacing is $0.5\lambda/\sin\alpha$.	The fringe spacing is $0.5\lambda/\sin\alpha$.
When $2\alpha = 0$ the bands will either be uniformly striped or totally black.	When $2\alpha = 0$ the two beams will add uniformly either constructively or destructively.
When $2\alpha = 180°$ the bands will produce a beating effect when moving with identical but opposite velocity.	When $2\alpha = 180°$ there will be standing waves, the spacing between them being 0.5λ.

This section has looked at the analogy between the moiré pattern and the interference effect of two intersecting mutually coherent beams of parallel light. A later section will examine the analogy between the moiré pattern and the interference patterns formed by the divergent light from two close point sources of mutually coherent light. In that case the wavefronts will be spherical and the interference fringes will consist of surfaces in space that are neither flat nor parallel. The new concept of "beams of coherent observation" will also be studied later to see how this can be used to calculate the shape of these three-dimensional interference surfaces and thus study how interference fringes are formed when they intersect objects.

1.2 Interferometers

1.2.1 Young's fringes

One of the most fundamental interference phenomena is caused by two simple pinholes (at A and B in Fig. 1.7) through which mutually coherent light is transmitted. The two sets of spherical waves that are emitted from the holes produce a fringe pattern in the form of hyperboloids with A and B as focal points. The hyperboloids are the loci of points that have a constant difference in distances to A and B and therefore have constant phase

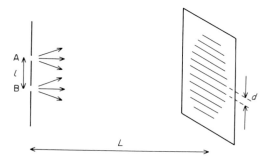

Fig. 1.7. Two pinholes in an opaque screen are used as point sources of light, which illuminate another screen at the right on which fringes are formed. These fringes are referred to as Young's fringes and their spacing (d) is calculated $d = k * \lambda/2$.

difference. This experiment was first described by Young and the fringes are usually referred to as Young's fringes. The separation (d) of the hyperboloids representing intersecting interference surfaces is given by $k * \lambda/2$, as described in Section 1.2, Fig. 1.5. Usually the distance l separating A and B is small compared to the distance L to the screen used for observation, and in that case the following approximation can be made:

$$d = k * \lambda/2 = \frac{\lambda}{2\sin\alpha} \simeq \frac{\lambda L}{l}. \tag{1.3}$$

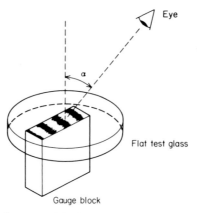

Fig. 1.8. The flatness of a gauge block is measured using a flat test glass placed in contact with the surface to be studied. Fringes are formed that, like the contour lines of a map, represent areas of constant height. The sensitivity of the device depends on the angle of observation.

1.2.2 Newton's rings

Another simple interference effect is referred to as "Newton's rings"; these are seen when two reflecting surfaces are separated by only a few wavelengths. One example is the coloured rings formed when oil is spilled on water. A practical use of this phenomenon is when the flatness of gauge blocks is measured using a plane test glass placed in contact with the surface to be measured (see Fig. 1.8).

Only specular (mirror-like) reflections can be studied by this method, therefore the angle of observation and the angle of illumination are identical. If observation is carried out perpendicular to the glass surface ($\alpha = 0$), then each of the fringes represents a separation of 0.5λ between object and

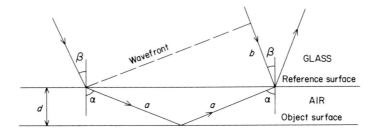

Fig. 1.9. Detailed diagram showing the situation illustrated in Fig. 1.8. The illuminating light is split into two components by the lower surface of the test glass. One component (the reference beam) is reflected by the glass surface, the other by the object surface under study. Interference occurs when the two beams recombine. The separation of the fringes represents a height difference of $k\lambda/2$. (See also Fig. 1.6.)

glass surface. For all other values of the angle of observation the sensitivity will be lower. In Fig. 1.9 the influence of oblique angles of illumination and observation is examined.

Two parallel light rays arrive at the lower surface of the glass plate (the reference mirror). The angle between these light rays and the normal to the surface is β inside the glass and α after it has passed into the air separating the glass from the surface of the object. The refractive index of the glass is n and the distance between the reference surface and the object surface is d. Snell's law of refraction gives the relation

$$\sin \alpha = n \sin \beta. \tag{1.4}$$

Now the number of wavelengths (m_{2a}) contained in the distance $2a$ is given by

$$m_{2a} = \frac{2d}{\lambda \cos \alpha} \tag{1.5}$$

and the number of wavelengths (m_b) contained in the distance b by

$$m_b = \frac{n2d}{\lambda n \cos \alpha} \sin^2 \alpha. \tag{1.6}$$

The number of fringes (m) corresponding to the distance d is equal to $m_{2a} - m_b$. Thus

$$d = \frac{m\lambda}{2 \cos \alpha} = mk\lambda/2. \tag{1.7}$$

Thus it is found that the distance d corresponding to one interference fringe is identical to the separation of the moiré fringes of Fig. 1.5 in Section 1.2. In comparing the present situation to Fig. 1.5, A is the light source while B is the point of observation made sensitive to the phase by receiving a reference beam (the reflection from the plane glass). The resolution of the interferometric system is represented by the separation of the interference fringes caused by one point of illumination (A) and one point of observation (B). The refractive index of glass has no influence on the calculation of the fringe separation in spite of the fact that the reference beam travels in glass all the time, whereas the object beam crosses an air film. The only important factors are the angles of illumination and observation at the object surface.

If observation is made from a short distance the angle α will vary over the object surface and therefore the sensitivity will vary, so that the interference surfaces will be neither equidistant nor plane. The intersection of these curved surfaces by a flat object produces a set of fringes in the form of circles (Newton's rings) on its surface.

If, however, the illumination and observation are collimated (made from infinite distances), then the interference surfaces become equidistant planes

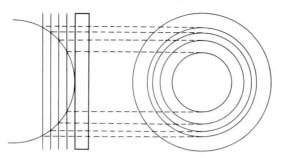

Fig. 1.10. A spherical surface in contact with the flat test glass of Fig. 1.8 produces a set of Newton's rings, each representing the object intersected by one of many interference surfaces, their separation being $k\lambda/2$.

that can be used for measurement of the flatness of, for example, metal surfaces. In this case one can simply assume that parallel to the plane glass surface there exist a number of flat interference surfaces separated by $k\lambda/2$, the intersections of which produce interference fringes on the object surface. Figure 1.10 illustrates how a spherical surface in contact with a plane glass produces a set of Newton's rings, each representing the object intersected by one of the interference surfaces.

The relevant experiments are all easy to perform. A plane piece of glass is pressed against a flat metal surface, e.g. a gauge block. If the surfaces are sufficiently clean that they come in close contact with each other it should be possible to see coloured fringes when diffuse daylight, e.g. from a cloud-covered sky, is reflected in the glass. The fringe representing the smallest separation will look almost black while the fringes representing increasing separation become more and more coloured and diffuse. Usually only about three or four fringes can be seen. However, if the light source is an ordinary mercury lamp, more than ten fringes can be resolved, and with a sodium lamp still more fringes can be seen. In the diffused light from a He–Ne laser more than 50 fringes can be counted.

In the last-mentioned case a convenient way to study the fringes is to let the laser beam be directly reflected by the glass and the object on to a screen. The laser should be placed at a distance of several metres from the object and its beam widened by a lens of short focal length, so that it covers the whole object surface. An image of the glass and the object is then projected on to the screen and the surface of the image is covered by high contrast fringes (Fig. 1.11).

1.2.3 The interferoscope

The interferoscope[1] is a special type of interferometer in which the katete of a prism is used as beam-splitter and reference surface instead of the plane

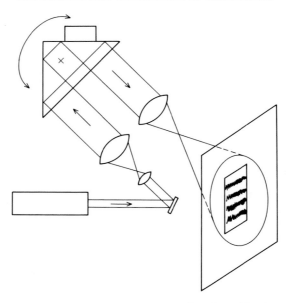

Fig. 1.11. The interferoscope. Light from the laser is reflected by the two katetes of a prism. The object is placed on top of the katete that is more or less horizontal. The light rays will strike its lowermost surface at an oblique angle which can be changed by tilting the prism. Thus the fringe pattern can be manipulated because the sensitivity to fringe formation will vary as described in Fig. 1.9.

parallel glass plate (Fig. 1.11). When a light beam is reflected an even number of times in a prism the direction of the reflected beam will be independent of small rotations of the prism. Therefore the light will reach the same place at the imaging lens when the prism angle is changed. The angle β (Fig. 1.9) is such that α is close to 90° (which represents total reflection). As a result, a small tilt of the prism produces a large difference in the angle (β) of illumination and observation. Tilting the prism c. 3° produces a change of no less than tenfold in the value of k.

Figures 1.12, 1.13 and 1.14 are three photographs taken through the instrument and show the interference patterns on a ground steel surface with some spherical depressions. In Fig. 1.12 the value of k is about 3, so that each fringe represents around 1 µm. Figure 1.14 demonstrates how a still more oblique illumination and observation angle lowers the resolution to about 5 µm per fringe, corresponding to a k-value of around 20. In Fig. 1.12 almost no surface features can be seen because of the high interferometric resolution, whereas Fig. 1.14 reveals spherical depressions the depth of which can be calculated by counting the circular fringes. The surface roughness is represented by the rugged horizontal interference fringes, the widths of which are about one quarter of a fringe separation or about 1.5 µm. With this

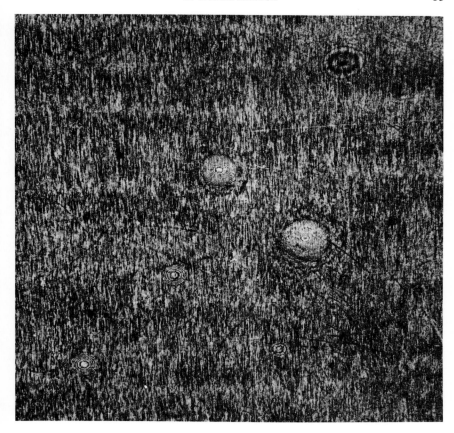

Fig. 1.12. Interferogram produced by the interferoscope shown in Fig. 1.11. One fringe corresponds to 1 μm. The object is a roughly ground steel surface with spherical impressions made by the impact of a steel ball. The area studied is about 20 mm × 20 mm. Because of the high sensitivity of the fringe formation combined with a rough surface very little information is found in this photograph. The randomness of the fringes caused by the microstructure of the surface means that its macroscopic shape cannot be compared with another similar surface either interferometrically or holographically.

instrument it is possible, using oblique illumination, to study interference fringes on so rough a surface that ordinary interferometers are useless.

By continuously changing the sensitivity it is possible to measure the depth of a depression even if its slopes are so steep that the fringes can never be counted. If, for example, a continuous change of the k-value by 10% causes the bottom of the depression to pass through three cycles of alternating brightness and darkness, then it can be seen that it is 30 surface interference separations deep.

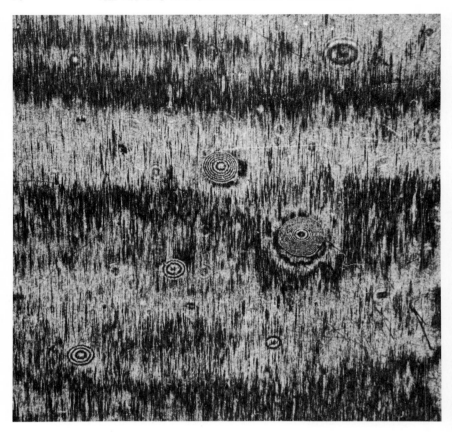

Fig. 1.13. The prism has been tilted so that one fringe corresponds to $2\,\mu m$. The object is identical to that of Fig. 1.12. As the broad horizontal fringes on the steel surface have separated it can be concluded that the maximum depth of the machining marks is just less than $2\,\mu m$.

Another strange feature of this type of interferometer is that it can produce black and white fringes even when it is illuminated by ordinary daylight. The reason is that the prism deflects blue light more than red light. Thus the angle α of the blue light will increase and so will its k-value.

As stated earlier, each fringe satisfies the relationship

$$d = nk\lambda/2. \tag{1.8}$$

Therefore, for a certain angle of illumination and for a certain type of glass the shorter wavelength is compensated for by the larger k-value, so that d will be equal for all wavelengths. As a result, the fringes caused by all colours of light will be localized at the same place and the interferometer produces high contrast fringes even in normal daylight (achromatic fringes).

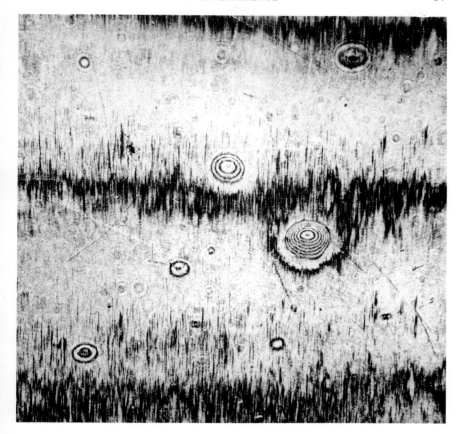

Fig. 1.14. One fringe corresponds to 5 μm. The object is still identical to that shown in Fig. 1.12, but the information available is significantly different. It is now easy to count the fringes down to the bottom of even the deepest impressions. The flatness of the surface and the surface roughness are easily studied. The directions of illumination and observation are so oblique that the *k*-value is around 15. In this case the macroscopic topography can easily be compared with that of another similar surface either by a conventional interferometer or by holographic interferometry.

This effect can easily be studied by placing a prism in contact with a shiny, flat, metal surface and observing the fringes directly with the naked eye. The fringes are seen just at the coloured border of total reflection.

1.2.4 The Michelson interferometer

The types of interferometry that have been described above all have the disadvantage that the reference surface (a reflecting transparent surface) has to be in contact with, or at least close to, the object surface. A number of

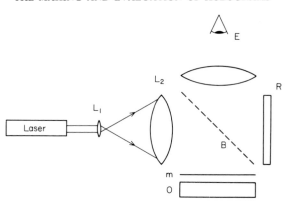

Fig. 1.15. A simplified drawing of the Michelson interferometer. In this instrument the surface under examination is not placed in contact with a test glass (reference surface) as in Figs 1.8, 1.10 and 1.11. Instead it is interferometrically compared to the mirror image of the reference surface. L_1 and L_2 are two lenses used to expand and collimate the laser beam. B is a beam-splitter, while O is the object surface and R the reference surface the mirror image of which (m) is in close contact with the surface of the object. The eye at the top studies the object with its fringes through a collimating lens.

interferometers have, however, been designed in which the object is compared, not with the reflecting reference surface itself, but with its mirror image.

One of the most important of these types of interferometers is the one invented by Michelson,[2] who, from about 1890, used it for many experiments that have been of great significance to physics. Of Michelson's experiments, perhaps the one which had the greatest influence on our view of the Universe was the "Michelson–Morley experiment" by which it was first proved that the measured speed of light was identical in all directions independently of the velocity of the Earth. This result put an end to earlier theories invoking the "ether" and a new era was opened in which the relativistic theories presented by Einstein have been of such enormous importance.

The Michelson interferometer is shown diagrammatically in Fig. 1.15. The beam from a laser is diverged by a lens of short focal length (L_1). Subsequently, the beam becomes sufficiently wide to illuminate the requisite area of the object; it is then collimated by a second, larger lens (L_2). The collimated light is then split by a semi-transparent mirror (the beam-splitter, B). One part of the light (the object beam) is reflected towards the object (O), from which it is reflected back through the beam-splitter to the eye (E) of the observer. The second part of the light from the laser, which passes directly through the beam-splitter, is reflected by the reference mirror (R) and thereafter is reflected by the beam-splitter and thus rejoins the object beam and travels towards the eye. When the object beam and the reference beam are reunited, interference fringes are formed. Because the illumination and the observation are made from directions perpendicular to the object surface, a deviation in the object surface of 0.5λ causes a phase change of 2π radians and thus one interference fringe.

Another way to explain the function of the instrument is to say that the observer sees the object surface with the mirror image (m) of the reference surface superimposed on it. The Michelson interferometer works in every way as if this image of the reference surface existed in the form of a flat glass on top of the object surface, but with the great advantage that this flat glass can be tilted and moved downwards so that it passes through the object. On each side of the image of the reference mirror there exist imaginary equidistant flat interference surfaces. The fringes seen on the object simply represent its intersection with these surfaces. By tilting the reference mirror the intersecting angles of the intersecting planes are changed to optimize the information about the topography of the object. By studying in which direction the centres of the circles of Fig. 1.16 move, it is possible to reveal whether the circles are caused by a concave (Fig. 1.16(b)) or a convex (Fig. 1.16(c)) surface.

Figure 1.15 shows that the interferometer has been so arranged that the object is illuminated by light that has been collimated by passing through a pinhole placed at the focus of lens L_1. The observed light is also collimated by the use of a lens. In this way the illumination and the observation angle will be constant over the whole object, which results in even illumination and also means that the sensitivity to fringe formation (and thus the k-value) will be constant.

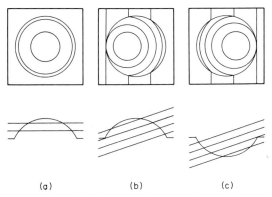

(a)   (b)   (c)

Fig. 1.16. A spherical surface like the one shown in Fig. 1.10 is studied through the Michelson interferometer illustrated in Fig. 1.15. By manipulating the interference fringes the sign of curvature, i.e. convexity or concavity, can be found. Tilting the reference mirror R of Fig. 1.15 causes the intersecting interference surfaces to be tilted in an analogous way. Similar fringe manipulation possibilities exist in hologram interferometry, as described in Chapter 6.

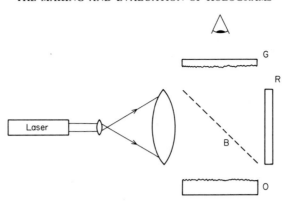

Fig. 1.17. The expanded and collimated laser light illuminates a Michelson interfero-meter, where B is the beam-splitter, R the reference mirror and O the object surface. The more or less parallel light rays will project an interference pattern on the ground glass diffuser, G. These fringes are real (or objective) as they are projected directly on to the screen. Had the surface of the object been a plane tilted mirror they would be analogous to the Young's fringes of Fig. 1.7.

Instead of looking directly at the fringe system, however, it is possible to let the light illuminate a diffuser (G) after it has passed through the inter-ferometer (Fig. 1.17). Because the light rays travel along straight lines, they will project on the screen an image of the object, together with its fringes. This type of fringe pattern is real (or objective) as they are directly projected on to a surface. If the object is a flat, tilted mirror, then from the diffuser one would see two point-like light sources. Thus the fringes are analogous to Young's fringes projected on to the screen.

However, an alternative procedure would be to turn everything the other way round and to illuminate the diffuse screen with coherent light while looking through the pinhole (Fig. 1.18). In this case exactly the same fringe pattern would be observed as before, but now it is influenced by the point of observation. If the pinhole is moved around the fringes will change. By using this method the sign of the slope can be found and the fringe spacing can be selected to suit the observation. This type of fringe will be referred to below as "interference fringes of observation" (or subjective fringes). From the screen two points of observation are visible and thus the fringes are analogous to Young's fringes of observation.

1.2.5 Interferometric measurement of differences

In whatever way the interferometer is illuminated and observed, it can always be used to measure differences between the object's surface and the reference surface. If, rather than being flat, the object is in fact spherical, it is possible

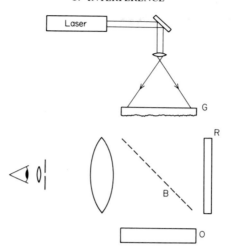

Fig. 1.18. By interchanging the point of observation and the point of illumination, the real interference pattern of Fig. 1.17 is replaced by "subjective fringes" or "interference fringes of observations". These fringes cannot be directly projected on to a screen. Their pattern depends on the point of observation, or rather, in the set-up described, on the position of the pinhole in front of the eye. This type of "subjective fringe" is common in hologram interferometry.

to compare it with a spherical reference surface so that the absence of fringes represents identical shapes.

It is, of course, also possible to compare the surfaces of two objects by photographing the fringe pattern formed when each object is compared to the same reference surface. If the two negatives are placed one on top of the other a moiré fringe pattern is formed that represents the difference between the two interference patterns.

To make this method work well, each of the negatives should record closely spaced fringes, obtained, for example, by tilting the imaginary interference surfaces in relation to the object surface. This tilt should be larger than the largest angle of any slope on the two object surfaces. When the two negatives are placed on top of each other a moiré pattern is formed that represents the difference between the two interferograms; if the reference mirror position was unchanged, it will thus represent the difference between the topographies of the surfaces of the two objects (Fig. 1.19). If one of the surfaces under comparison is flat, its fringes will be in the form of straight lines corresponding to the tilt angle. Thus it is possible to eliminate the tilt of a deformed and tilted surface simply by adding a grid of straight fringes on the photographic image of the interferogram of the surface.

Thus there exist two possible methods for the comparison of the surfaces of two objects with one another. One way is by placing the surfaces in each

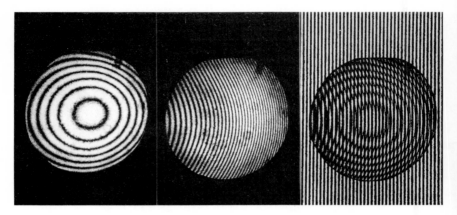

Fig. 1.19. It has already been shown in Fig. 1.16 that the fringe pattern can be manipulated by tilting the reference mirror of a Michelson interferometer. Analogous manipulation can be carried out in the image plane by placing a grid on top of the fringe-covered image. To the left is shown the spherical object surface covered by circular fringes. In the middle the reference surface has been greatly tilted. To the right is shown how placing a grid on top of the middle image produces a moiré effect, eliminating the influence of the tilt of the reference mirror. Similar fringe manipulation methods exist in hologram interferometry (see Figs 6.11–6.15) either by interference at the hologram plate (e.g. sandwich holography) or by moiré effects in the image plane.

arm of an interferometer: tilting one of the surfaces makes it possible to tilt the intersecting interference surfaces. The other way is to compare the surfaces of the two objects by exposing two negatives through an interferometer with its reference mirror unchanged and thereby producing the moiré pattern of the two processed negatives. By adding a grid pattern to one of the negatives it is possible to simulate a tilt of the intersecting surfaces. One advantage of the latter method is that this change can be made long after the object has been removed from the interferometer.

Interferometry combined with the use of moiré patterns is very similar to holography. However, before starting on our main subject, let us study some more basic optical principles.

Any two sets of fringes that combine to produce a new set of fringes are called *primary fringes* in this book and the new fringes are called *secondary fringes*. A combination of two sets of secondary fringes can also produce fringes, i.e. tertiary fringes, etc. The higher order fringes represent the difference of two lower order fringe patterns. Thus the term primary fringes refers, for example, to the stripes that represent the wavefront of our moiré pattern analogy and also to the wavefront itself. The term secondary fringes represents the moiré fringes caused by the two striped patterns; it also represents the interference pattern caused by the two coherent light beams.

The moiré fringes produced by placing two transparencies of interference patterns on top of each other are tertiary fringes, etc. As both moiré and interference patterns are made up by using secondary fringes, it is no wonder that these two phenomena are analogous in the way described earlier. In this book much of the reasoning will be based on this type of analogy.

1.3 Energy balance of interference phenomena

1.3.1 Intensity amplification

It has already been mentioned that if two mutually coherent beams from different angles illuminate a screen there will be fringes of alternating brightness and darkness, that the sum of the energies will be conserved, and therefore that the local brightness of the light areas will be higher than the arithmetic sum of the brightnesses of the two beams.

If there were no dark fringes the intensity would be the sum of the intensities of the two beams. If the dark fringes occupy half the illuminated area, then the other half has to have double the intensity of the sum of the two beams. Thus the light intensity at some points would be four times as high with two beams as it would be with only one beam. From this reasoning it can be seen that interferometry is a non-linear process. Light intensity is defined as the square of the amplitude, and therefore an argument based on amplitudes will arrive at the same result.

If this phenomenon is scrutinized more carefully, it is found that the brightness (measured in watts per square metre) represents the square of the amplitude of the electromagnetic radiation and that these amplitudes are added arithmetically. Thus, if the two intensities i_1 and i_2 are added coherently, the resulting intensity will be $(\sqrt{i_1} \pm \sqrt{i_2})^2$ (taking the positive values of the roots). Figure 1.20 illustrates the result of adding together the intensities of two coherent light beams which are at an angle to one another and have different intensities. The intensity of one beam (the signal beam, i_s) is constant at 1 mW while the intensity of the other beam (the reference beam, i_r) is varied from 1 mW to 20 mW. Surprisingly, the modulation of the resulting fringes (the difference in intensities between bright and dark fringes) increases rapidly as the strength of the reference beam increases. In spite of the fact that the signal beam is only 1 mW it will produce a modulation of no less than 18 mW if it is added to a reference beam of 25 mW. To gain this impressive result, however, it is necessary to accept a background intensity of 12 mW. This result can, however, be tolerated if the signal-to-noise ratio can be kept at a reasonable level. This is possible because the reference beam can be made very clean, especially as it is a local beam that never has to go out of the closed system and therefore can be kept free of spatial and temporal noise.

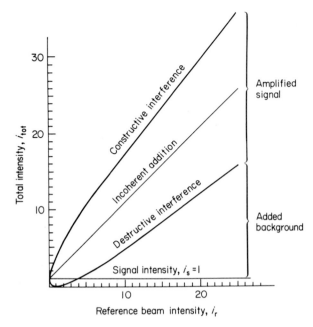

Fig. 1.20. The addition of the reference beam to the object beam (signal beam) in an interferometer produces a non-linear effect that can be used for intensity amplification. Let the signal beam have an intensity of one unit while the added reference beam has an intensity that is varied from 0 to 24 units. Their incoherent addition will result in a straight line of intensity from 1 to 25 units. Constructive interference, on the other hand, results in an intensity of at most 36 units. Thus switching the signal beam of one unit on and off results in an amplified intensity signal that varies by no less than 12 units. The amplifying power of the reference beam is of no small importance in holography.

Thus, by simply adding a strong coherent reference beam, the intensity modulation caused by phase changes can be amplified in such a way that it becomes many times larger than the signal beam that causes the modulation. There will even be an amplification if the phase is kept fixed but the intensity of the signal beam is varied. If, for example, the 1 mW signal beam is switched on and off repetitively, there will be a maximal outgoing variation in the interference pattern of no less than 9 mW when the power of the reference beam is 20 mW. However, in both cases it is only the absolute value of the modulation that is amplified, not the ratio between the modulated (a.c.) and the non-modulated (d.c.) signal.

1.3.2 Energy conservation in interferometry

How real are the interference fringes? Do they really exist by themselves or are they not formed until they illuminate a surface? Many experiments have proved that two intersecting light beams have no influence whatsoever on each other if there is no matter within the intersecting area. Therefore, it is interesting to consider what happens when matter is introduced into the interference pattern.

Let us re-examine Fig. 1.5, in which two laser beams travel from A to D and from B to C, respectively. If a small particle is passed through the fringe system where the two beams intersect, it will be illuminated when it is in the bright fringes and it will not be illuminated when it is in the dark fringes. Thus, the particle will alternately appear bright and dark. Its scattered light can be observed by a detector placed, for example, between C and D. If the particle moves parallel to the fringes, the illumination will be uniform and no pulses will reach the detector. If it moves normal to the fringes, the pulse frequency will be a measure of its speed in the following way:

$$v = f k \lambda / 2 = \frac{f \lambda}{2 \sin \alpha}, \tag{1.9}$$

where v is the velocity (in metres per second), f is the frequency (in megahertz), λ is the wavelength of light (in micrometres), and α is the half-angle between the two beams.

Instruments based on this principle are called laser velocimeters and are used to measure the speeds of small particles in gases and liquids. Another way to calculate the speed from the beat frequency would be to study the doppler shift of the frequency of the light scattered by the particle. The particle moves towards the light waves of one of the laser beams with speed $v \sin \alpha$. Thus, the frequency of this light will appear to the particle to be $(v/\lambda) \sin \alpha$ higher than if the particle was at rest. The other laser beam will cause an equal doppler shift, but the frequency will instead appear to be lower. The two doppler-shifted frequencies are scattered towards the detector placed at the bisector of the angle of intersection of CB and DA. The difference in light frequency will produce a beat signal having the frequency

$$f = v \frac{2 \sin \alpha}{\lambda}. \tag{1.10}$$

Thus, the use of doppler shift produces exactly the same result as the use of interference fringes. However, the latter method is to be preferred because of its simplicity, its generality, and, last but not least, the philosophy upon which it is based. The detector placed at the bisector of the angle of intersection of CB and DA (Fig. 1.5) receives pulses simply because *objects that are not illuminated have no influence on light.* Only when the particle is within a bright fringe will it scatter light to the detector. When the particle is in a dark fringe it cannot be detected at all because it emits no information. It is not even possible to prove that the particle still exits. Thus, one cannot in a true sense speak about a doppler shift between the two light beams that are scattered by the particle.

What if the detector is placed at C or D—will it be possible to detect the shadow of a particle even if it is situated in a dark fringe? The answer is no. If no energy arrives at the particle, it can produce no information. If, however, the object is in a bright fringe, shadows are cast at both C and D. Thus the velocity of particles passing through the fringe system can be measured by detectors at C or D. Laser velocimeters based on this method have been built, their main disadvantage being a low signal-to-noise ratio because the scattered light is weak compared to the direct beam.

If the illumination rule is true, a thin, flat, opaque surface can be placed along one fringe; surprisingly, it is then observed that light passes through that surface. The rule predicts that nothing will happen if an object is introduced into a dark fringe. This author tried the experiment and found, as would be expected, that nothing happened. When the surface was placed parallel to the fringes, light from A and B reached C and D and it was impossible to decide whether the surface was penetrated by the light or whether it was simply reflecting the light beams. If a mirror is placed in an interference fringe, it will reflect light from A to C exactly as if the light came from B because the angle of incidence is always equal to the angle of reflection. A rougher surface was then tried which was not such a good mirror. The light beam at C then disappeared. Had the illumination rule been proven wrong? No! Because the surface was made rougher some of its peaks reached beyond the darkness into a bright fringe, and thus the rule no longer applied. Finally, the fringes were broadened by decreasing the angle α, and at a certain value the beam at C reappeared. This is a well-known phenomenon, viz. a rougher surface appears more mirror-like as it is illuminated from a more oblique angle (see Section 1.3.3). So once again the illumination rule appeared to be true.

What will happen if a semi-transparent mirror is used? Experiments show that if it is placed with all of its surface within one dark interference fringe nothing changes at C or D. If the mirror is kept parallel to the fringes but moved out into the brightness of one fringe, all the light will be directed either to C or to D. If the mirror is placed at an angle to the fringes so that it intersects, for example, five fringes, then five fringes will also appear at both C and D. Dark fringes at C correspond to bright fringes at D because the energy not reaching C will reach D instead. Thus, there exist at least two methods to predict the fringes at C and D: the conventional way, calculating the angle between the light passing through the mirror and the light reflected by the mirror and then finding fringe separation using equation (1.3), and another way, using the equation first and then counting the number of inter-sections between the mirror surface and the interference fringes. In this book the author intends to demonstrate the advantages of using the latter method, especially when the mirror is replaced by an object on which measurements are to be made.

One final, but not too serious, look will now be taken at interference fringes. What is a fringe pattern: dark fringes on a bright background or bright fringes on a dark background? Perhaps the answer can be found in the following way. If one of the light sources, for example A of Fig. 1.5, is moved to the right with a speed of one dark line per second, the moiré fringes would move downwards with a speed of one fringe per second. The distance between moiré fringes is much larger than the distance between wavefronts (the lines at A) and, therefore, the moiré fringe speed is much higher than the speed of the source A. If A is moved with one-hundredth of the speed of light, and if the distance between the interference fringes is 1000 wavelengths, the fringes would move with ten times the speed of light. But Einstein states that nothing can move with a speed exceeding the speed of light. If he is right, the fringes have to be "nothing", and consequently, the fringe pattern has to be dark on a bright fixed background! The same reasoning applies to speckle patterns, which consequently consist of dark spots.

Those with a sound knowledge of physics will probably say that the above statement is nonsense because interference fringes move not with the group velocity c but with a phase velocity which may very well exceed c. The answer to this objection is that the existence of light is necessary for the phase velocity to form and thus no part of the statement is invalid.

1.3.3 Diffraction-limited resolution

Finally, if two laser beams produce bright and dark interference fringes, what would happen if the two beams were almost parallel to each other and so narrow that only one dark fringe was produced within the intersection? Could the two beams end up in total darkness? In that case what would happen to the energy, or how would the apparent contradiction of the law of conservation of energy be prevented?

Let us study the experiment in Fig. 1.5 again. To make the fringes as broad as possible α must be made as small as possible. Making A and B closer together will reduce α and if A and B are two windows, lenses, or mirrors the closest they can be to each other is in contact. Increasing the distance (L) between A and B and the intersection will broaden the fringes. If L is made sufficiently long, one fringe will become broader than the diameters of the beams, and at that point the light energy will have just disappeared. However, this contradiction of nature cannot arise and therefore the beams must widen with distance L. The smallest possible diameter at the intersection has to be

$$D = k\lambda/2 = \frac{\lambda}{2 \sin \alpha} \simeq \frac{\lambda}{2d/2L} = \frac{\lambda L}{d}, \tag{1.11}$$

where D is the beam diameter at the intersection, d is the beam diameter at

A and B, λ is the wavelength, and L is the distance from A and B to the intersection. This equation represents the accepted estimate of the smallest possible divergence of a laser beam. It is said to represent the diffraction-limited divergence.

However, could the beams not be concentrated with the use of lenses in such a way that they were narrower than one interference fringe? Returning to Fig. 1.5, consider A and B to be two lenses that concentrate two beams so that they intersect at the focal points of the lenses. One interference fringe (dark + bright) will have the width k at the intersection.† If the beam is focused to a diameter less than k, the energy would disappear. The closest the lenses can be is in contact with one another. In that case, the distance between their centres would be equal to the lens diameter, and hence we would have $\alpha = \beta$. No energy would disappear if a lens could not focus to a spot smaller than k. Thus the smallest possible diameter at the focal point has to be

$$D = \frac{\lambda}{2 \sin \alpha}. \tag{1.12}$$

The smallest spot to which a lens can focus is given in the literature as

$$D = \frac{1.21\lambda}{2 \sin \alpha}. \tag{1.13}$$

This value, which represents the resolution of a lens, is often referred to as Abbé's limit of resolution.[3] The discrepancy of 1.21 found when a circular illumination or observation area such as a lens is involved is caused by the fact that the total area of the lens, not just the two diametrically opposed points, is used. In order to produce one single image, some of the resolution has to be sacrificed. The "holo-diagram" represents the fundamental resolution of any optical system, and thus is a tool that concentrates a large amount of information into one single method.

1.4. Interference from two sets of spherical waves

Up to this point we have concentrated on the moiré pattern analogy to the interference effect of two intersecting coherent beams of parallel light. In this section the moiré analogy to the interference patterns formed by the divergent light from two sources of coherent light will be examined. In this case the wavefronts will be spherical and the interference fringe surfaces in space will not be flat or parallel or equidistant. Utilization of the new concept of "beams of observation" will be studied as a method for calculating the

---

† It has been proven in the literature that a focused converging beam is parallel for a short distance about the focal point.

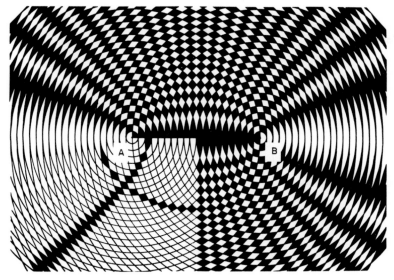

Fig. 1.21. A and B are the centres of two sets of concentric circles in a bipolar coordinate system. The moiré fringes form one set of ellipses and one set of hyperbolas. To emphasize these patterns every second rhomboid area has been painted black except for one-quarter of the diagram where just one single ellipse and hyperbola have been marked.

shape of these three-dimensional interference surfaces and in order to understand how interference fringes are formed when they intersect objects.

1.4.1 The holo-diagram

Consider the fringes formed by the spherical wavefronts from two mutually coherent point sources of light (A and B of Fig. 1.21). If the light waves are in phase at the two sources, then they will also be in phase at all places in space at which the difference in distances to the two sources is zero or a whole number of wavelengths. The foci of these points will be a set of rotationally symmetric hyperboloids, their focal points being A and B respectively. Thus, along these hyperboloids the interference will be constructive; it will be destructive at intermediate positions.[4]

In the space around A and B there will exist interference surfaces in the form of alternating bright and dark hyperboloids. These fringe surfaces can, of course, not be seen directly in empty space but only when they are intersected by matter, e.g. by a flat, white surface.

Figure 1.21 demonstrates the two-dimensional moiré pattern analogy to the interference of two spherical wave systems. A and B are the centres of two sets of circles in a bipolar coordinate system. The radii of the circles around

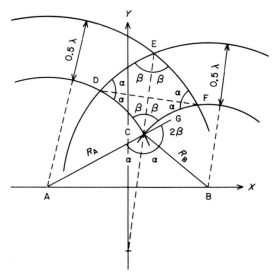

Fig. 1.22. The angles of one of the rhombs of Fig. 1.21 are studied and the two diagonals calculated. The statements of the figure are true only if λ is infinitesimal.

A and B are labelled R_A and R_B respectively. The separation between two adjacent concentric circles is 0.5λ.

If the separation is infinitesimal, the intersections of the two sets of circles will form a set of rhombs which form one set of ellipses $(R_A + R_B = n\lambda)$ and one set of hyperbolas $(R_A - R_B = n\lambda)$. In Fig. 1.21 every second rhomb has been blackened except in one quadrant where only two chains of rhombs have been filled in to emphasize the ellipses and the hyperbolas.

From Fig. 1.22 it is possible to show that the length of the diagonal CE (the separation of the ellipses) is $0.5\lambda/\cos\alpha$ and that the diagonal DF (the separation of the hyperbolas) is $0.5\lambda/\sin\alpha$. The angle between R_A and R_B is 2α and k and k^* are defined as $1/\cos\alpha$ and $1/\sin\alpha$ respectively. The angle α is constant along the periphery of a circle through A and B and therefore it is possible to produce Fig. 1.23, in which the arcs of circles represent loci of constant separation between the ellipses (k-value without parentheses) and the hyperbolas (k^*-value within parentheses).

This "holo-diagram" was introduced in 1968 as a practical device for the production and evaluation of holograms. However, only the ellipses and the k-value were used for that purpose. Figures 1.22 and 1.23 represent cross-sections of a three-dimensional body produced by rotation of the fringes around the AB axis. In the three-dimensional case the circles are transformed into spheres, the ellipses into ellipsoids, the hyperbolas into hyperboloids and the k-circles into toroids.

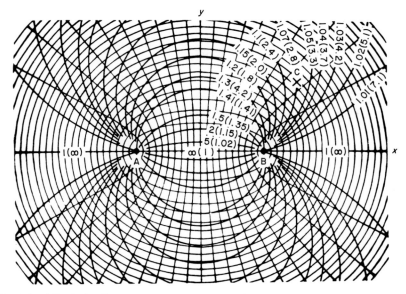

Fig. 1.23. The holo-diagram based on one set of ellipses and one set of hyperbolas with common focal points as described in Figs 1.21 and 1.22. The arcs of circles represent the loci of constant separation between the ellipses (k-value without parentheses) and the hyperbolas (k^*-value within parentheses). The hyperbolas represent Young's fringes or the diffraction-limited resolution of Abbé if the foci are two light sources or two observation points respectively. The ellipses represent imaginary interference fringes or the "interference-limited" resolution of, for example, holographic interferometry. In that case the focal points represent one light source and one point of illumination, respectively.

These surfaces have many properties of interest. The hyperboloids remain stationary in space if the circles around A and B represent spherical wavefronts of light that move outwards from two mutually coherent light sources. The ellipsoids, however, will move outwards with a velocity (v) that is higher than the speed of light (c): $v = kc$. Thus the hyperboloids represent stationary interference surfaces which, when intersected by an object on its surface, produce the ordinary interference fringes caused by two divergent beams of coherent light.

The ellipsoids, on the other hand, will remain stationary in space if the circles around A represent spherical wavefronts that move outwards while those around B move inwards, or vice versa. The hyperboloids will in this case move in a transverse direction to the right with a velocity (v) that exceeds the speed of light: $v = k^*c$. Thus the ellipsoids represent stationary fringes caused by one divergent and one convergent beam of light.

If, finally, the light converges towards both A and B, the stationary interference surfaces will once again form a set of stationary hyperboloids.

Thus it has been shown that if the points A and B are two mutually coherent light sources then the interference surfaces are created in the form of rotational symmetric hyperboloids, their spacing being k^*. Let now the direction of light around B change through 180° so that the waves move towards B, for example, because B represents the focus of a lens or a mirror illuminated by light that is coherent with A. In that case the hyperboloids will be transformed into ellipsoids, which means that the fringes turn through 90° (the diagonals of a rhomb are perpendicular) at the same time as their separation changes from k^* to k, satisfying the equation

$$\left(\frac{1}{k}\right)^2 + \left(\frac{1}{k^*}\right)^2 = 1. \tag{1.14}$$

If the direction of light around A is changed too, so that the waves also move towards A, then the fringes will once more rotate through 90° and the ellipsoids turn into hyperboloids, their spacing again becoming k^*.

1.4.2 Wavefronts of observation

The fringes might become ellipsoids when A is a light source even if light is not focused towards B. Consider the following situation. If B is a point of observation, it will only react to light that is directed towards B because only the light that is used for the observation will influence its result. Thus the observation only reacts to spherical wavefronts moving towards B. If A illuminates a large white-painted surface, B will react as if this surface focused light towards B. (The light scattered in other directions is not observed.) If the observation at B is later made phase-sensitive (e.g. by also receiving a direct beam, a reference beam, from A), then B will only react to spherical waves moving towards B and having a certain phase. This is exactly the situation described earlier, when light was focused towards B and the fringe system could be evaluated as if this was the case.

It is therefore possible to introduce a set of spherical "wavefronts of observation" that are mutually coherent with the beam of illumination but move outwards from the point of observation. When this beam of observation intersects the beam of illumination interference surfaces are produced in the object space. These surfaces are identical to those that would be formed by ordinary light waves moving inwards, towards the lens. They are called imaginary fringe surfaces, because they cannot be seen directly on a diffuse screen, as can ordinary interference fringes.

Finally, if both A and B are points of observation, the fringes are said to be caused by interference surfaces of observation. These fringes form a set of hyperboloids, identical to those caused by two points of illumination. Thus they are referred to as Young's fringes of observation.

1.4.3 Real and virtual images

The hyperboloids and ellipsoids of Fig. 1.21 have many properties of interest.
It should, however, be remembered that the moiré patterns of Fig. 1.21 are
only analogies. The rhombs are caused by the fact that the sine waves of light
were approximated into square waves. The real shapes of the ellipsoids and
the hyperboloids are seen in Fig. 1.24(a) and (b). If they are recorded on
photographic material, they will have focusing effects both by reflection and
by diffraction. Every ellipsoid will reflect all the light from A towards B, thus
a real image of A is formed at B. It is said that this point B is reconstructed by A, which emits a reconstruction beam. All the light waves will
arrive in phase at B because all the path lengths are multiples of the
wavelength; it is said that the "Bragg condition" is fulfilled everywhere on the
surface of the ellipsoid. Any intersection through the ellipsoids will produce
a Fresnel zone-plate that, when illuminated from A, reconstructs B by
diffraction. This type of zone-plate is not restricted to being flat, it may have
any curvature.

Every hyperboloid will reflect light from A as if it came from B, thus a
virtual image of A is formed at B. In this case all the light waves will be
reflected as if they had been emitted in phase from B because the difference in
path length from the point of reflection towards A is a multiple of the
wavelength, thus the Bragg condition[5] is also fulfilled everywhere on the
surface of the hyperboloid. Any intersection through the hyperboloids will
produce a Fresnel zone-plate that, when illuminated from A, reconstructs B
by diffraction.

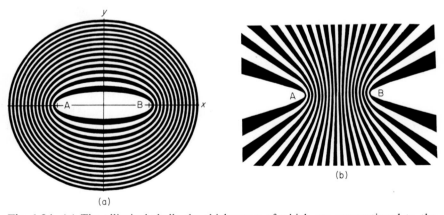

(a)   (b)

Fig. 1.24. (a) The elliptical shells the thicknesses of which are proportional to the
k-values of the diagram in Fig. 1.23. These shells represent imaginary interference
fringes or the "interference-limited" resolution of hologram interferometry. (b) The
hyperbolic shells the thicknesses of which are proportional to the k^*-values of the
diagram in Fig. 1.23. These shells represent Young's fringes or the diffraction-limited
resolution of Abbé.

If the direction of the light is changed before reconstruction so that, for example, the ellipsoids are illuminated by light focused towards A, it will be reflected (and diffracted) as if it came from B, and therefore a virtual image of A is seen at B. Finally, if light is focused towards A, it will be reflected (and diffracted) by the hyperboloids to B, and therefore a real image of A is formed at B.

One new way to sum up the results obtained above is by using the following rule for postulating whether a real or a virtual image will be reconstructed. Let divergent light be referred to as (-1), convergent light as $(+1)$, a virtual image as (-1) and a real image as $(+1)$. Thus, for example, two convergent beams produce a set of hyperboloids. If these hyperboloids are recorded on photographic material and later illuminated by a divergent reconstruction beam, the image is calculated in the following way:

$$(+1) \times (+1) \times (-1) = -1. \tag{1.15}$$

Thus the image is virtual.

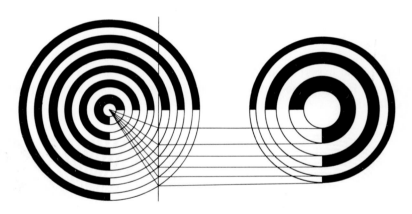

Fig. 1.25. A Fresnel zone-plate produced by letting a spherical surface be intersected by a number of equidistant parallel flat surfaces. By diffraction this lens focuses light from infinity to the centre of the spheres.

Fresnel zone-plates are usually described as being formed by one flat intersection of a set of spherical wavefronts, as seen in Fig. 1.25. However, this case is only one special example of our general principle based on the intersections of the moiré pattern formed by two sets of spherical wavefronts. The centre of one of the sets has been moved to infinity so that its wavefronts become flat. Therefore the ellipsoids and the hyperboloids are both transformed into paraboloids.

References

1. N. Abramson. The "interferoscope", a new type of interferometer with variable fringe separation. *Optik* **30**, 56 (1969).
2. G. Freier. *University Physics.* Meredith Publishing Company, New York (1965), p. 502.
3. M. Born and E. Wolf. *Principles of Optics.* Pergamon Press, New York (1965), p. 397.
4. N. Abramson. The holo-diagram. VI. Practical device in coherent optics. *Appl. Opt.* **11**, 2562 (1972).
5. G. Freier. *University Physics.* Meredith Publishing Company, New York (1965), p. 537.

2

The holographic process

Let us recapitulate from Chapter 1 what happens when two mutually coherent, collimated laser beams intersect at an angle so that they produce interference surfaces in space. The moiré analogy to this phenomenon was demonstrated in Fig. 1.5, in which the two beams are represented by bands of striped patterns that stand for the wavefronts.

The moiré pattern of Fig. 1.5 is only a two-dimensional representation of the three-dimensional case in which interference surfaces exist in space. These surfaces, which bisect the angle between the two beams, are equidistant and perpendicular to the plane of the paper of the figure. Their separation is k^*, as defined in Section 1.4.1.

2.1 Recording by two intersecting laser beams

If a photographic plate is placed within the region where the two beams intersect, a cross-section of the interference surfaces will be recorded. After exposure the plate is developed, fixed and re-illuminated by one of the two laser beams. The recorded dark fringes will of course throw shadows in the beam and, as the light that has passed through the plate cannot distinguish these shadows from interference fringes, it reacts as if it had arrived in the form of two beams; accordingly light leaves the plate in the form of two beams having exactly that angle that would have produced the recorded fringe pattern. Thus, this pattern is a memory of the angle separating the beams and this memory is read by shining one beam through the plate. This photographic plate with its fringe pattern is a grating that, by diffraction, deflects the light from one beam so that the other is reconstructed. The grating is the simplest form of a hologram.

2.1.1 True and conjugate beam deflections

From Fig. 2.1(a) and (b) it can be seen that identical fringe patterns are produced on a thin emulsion at the holographic plate (H) independent of whether the object beam (O) arrives from the left or from the right of the reference beam (R) as long as the angle 2α is the same. When the plate is processed and reconstructed by the reconstruction beam (R) there is therefore no information about the sign of the angle 2α and accordingly two beams are produced, one to the left and one to the right, as seen in Fig. 2.1(c).

However, if the photographic emulsion is thick, it will record not only a

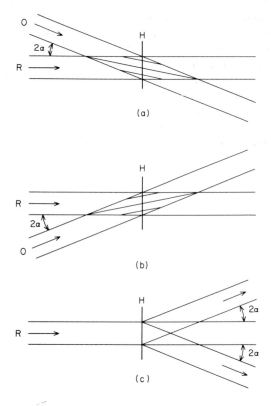

Fig. 2.1. The object beam (O) produces the same fringe pattern on the hologram plate whether it intersects the horizontal interference beam (R) from above (a) or from underneath (b). The reason is that the fringe spacing only depends on the absolute value of the angle 2α, not on its sign. When the processed plate (H) from either case (a) or case (b) is again illuminated by a horizontal beam both the object beams will be reconstructed because the plate does not remember the sign of the angle 2α. This statement is true only for a thin emulsion. A thick emulsion on the plate would also remember the direction (the bisector of 2α) of the interference surfaces.

cross-section of the fringe pattern (their separation along the plate) but also their three-dimensional shape (their direction). As is evident, there is a difference in the angles of the moiré patterns of Fig. 2.1(a) and (b), thus the information about the sign of 2α is recorded. Therefore a thick hologram, or volume hologram, has the ability to reconstruct the true object beam exclusively without producing any false beams.

The interference surfaces recorded in the thick emulsion can be thought of as being effective in three different ways. First, they diffract light just like a thin emulsion, and it is this effect that determines the angle of the reconstructed beam. Second, they reflect light in exactly the direction of diffraction, because they bisect the angle 2α and their separation is such that the Bragg condition (see Section 1.4.3) is fulfilled. Finally, they act like a venetian blind that prevents light coming through which would otherwise have produced false beams. Because of the added mirror effect the efficiency of a thick hologram can be much higher than that of a thin hologram.

One problem, using a thick hologram, is that its efficiency can decrease considerably if the angle of the interference surfaces recorded in the emulsion changes between exposure and reconstruction. The mirror effect will be in the wrong direction and the venetian blind effect might even shut off the true beam that should be reconstructed by diffraction. One reason for this change

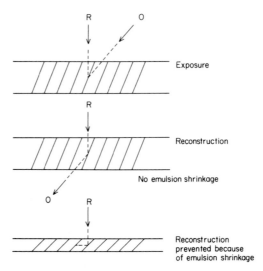

Fig. 2.2. The interference surfaces intersecting a thick emulsion produce sheets of exposed grains. If these sheets are unchanged during reconstruction, they will work like mirrors, enhancing the diffraction process. If this angle is changed during the processing of the plate, they might lower the efficiency of a hologram or even prevent the reconstruction altogether because of a venetian blind effect.

in angle of the recorded interference surfaces could be emulsion shrinkage caused by the processing of the plate, as seen in Fig. 2.2.

However, if the object beam and the reference beam illuminate the plate from such angles that their bisector is perpendicular to the plate, then the interference surfaces will stay perpendicular to the plate independent of any changes in the emulsion thickness. Thus, the plate angle should, if possible, be so arranged that this condition is fulfilled.

2.1.2 Other false beams

Let us now go back to the thin emulsion plate again and see if it is possible to reconstruct more false beams than the one just described. Do there exist still more object beams that, when combined with the fixed reference beam (R), produce the same fringe frequency as the object beams of Fig. 2.1? Yes, an object beam moving in the opposite direction would produce identical fringe spacing at the plate surface.

Consider Fig. 2.3(a) in which a hologram plate (H) is positioned at an angle γ to the bisector of the reference beam (R) and the object beam (O_1) which intersect at the angle 2α. The fringe spacing (f_1) on the plate will be

$$f_1 = \frac{\lambda}{2 \sin \alpha \cos \gamma}. \tag{2.1}$$

Figure 2.3(b) describes the situation when another object beam (O_2) enters the plate from the side that is opposite to that from which the reference beam enters. The reference beam (R) remains as in Fig. 2.3(a). The new object beam (O_2) is directed so that it coincides with the mirror image of the object beam (O_1) which is identical to the one in Fig. 2.3(a).

As before, the hologram plate is tilted by an angle γ with respect to the bisector of R and O_1 (which intersect at the angle 2α). Thus the beam O_1 will be at an angle $\gamma - \alpha$ relative to the normal to the plate. The angle between R and the beam O_2 (which is O_1 reflected by the plate) will therefore be given by

$$\alpha + \alpha + (\gamma - \alpha) + (\gamma - \alpha) = 2\gamma.$$

If the two beams R and O_2 interfered they would produce interference surfaces in space separated by the distance d_2:

$$d_2 = \frac{\lambda}{2 \cos \gamma}. \tag{2.2}$$

The angle between the fringe surfaces and the hologram plate would then be α (Fig. 2.3(b)). Thus the fringe spacing (f_2) on the plate would be

$$f_2 = \frac{\lambda}{2 \cos \gamma \sin \alpha}. \tag{2.3}$$

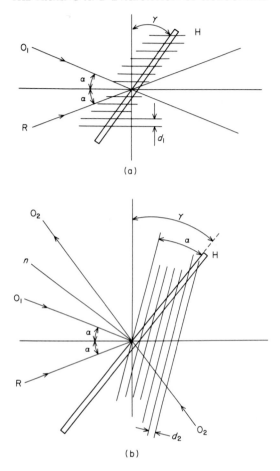

(a)

(b)

Fig. 2.3. (a) When the object beam (O) and the reference beam (R) illuminate the plate from the same side interference surfaces will intersect the plate in a direction bisecting the angle 2α which separates O_1 and R. (b) Had the object beam arrived not from the object (O_1) but from its mirror image (O_2) the interference surfaces would intersect the angle separating O_2 and R. It is proved that the intersections of the plate surface by the interference surfaces will be the same for the situation of (a) as for (b). Thus the image seen in the reflection of an ordinary hologram is just as true as the ordinary true image seen in transmission.

The result that $f_2 = f_1$ proves that a true virtual image will be reconstructed at the mirror image of the object when this is reflected by the emulsion side of the plate. As usual, there will also be a conjugate image on the other side of the reflected reconstruction beam.

Finally, the plate will also reconstruct two object beams if it is illuminated by a reconstruction beam that is antiparallel (moving in the opposite direction) to the former reference beam. The reason for this is that the pattern recorded in the thin emulsion is two-dimensional and thus identical (but mirror reversed) if the plate is turned round. Therefore, if a thin emulsion is used, six different reconstructed beams will exist, only one of which is a true replica of the original object beam.

There is another reason for the plate to reconstruct still more false beams. If the darkness of the fringes is not recorded by the photographic emulsion in a faithful way, new beams will be formed in directions separated from the reference beam by multiples of 2α. In the fringe pattern, the amplitude of the electromagnetic radiation will vary as a sine function. The intensity will thus vary as the square of the sine function. The photographic emulsion reacts to the intensity and thus records the sine-squared function. During reconstruction the intensity variation, caused by the varying transmission of the plate, will produce an amplitude variation in the form of a sine function. As a sine function cannot be separated into different Fourier components, just one frequency will be reconstructed corresponding to one beam direction.

However, if the recording process is not faithful but, because of saturation say, records a sine function as a square function, then the hologram reconstructs beams in such directions that they, if moving in opposite directions, would together produce that square function. Together with the reconstruction beam each of those beams would produce a fringe pattern that exists in the Fourier spectrum of the square wave. However, when ordinary photographic emulsions are exposed and processed in the recommended way, the darkening of the emulsion will be a linear function of the exposure intensity such that the effect of these higher-order diffraction beams will be negligible.

One strange feature of the photographic material (the holographic plate) is that it can record and reconstruct an extremely high number of object beams in different directions without mixing the information. It is as if during reconstruction it was possible to unscramble scrambled eggs! The only limitation to this strange process is that the sum of the intensities of the object beams should nowhere be higher than the intensity of the reference beam. If that should happen the beams start to have an influence on each other, i.e. there is cross-talk between the different beams.

Therefore, the reference beam should be made at least a few times stronger than the total power of all the object beams. In that case the hologram plate may be covered by many different fringe patterns superimposed on each other, having different angles and different frequencies. During reconstruction each beam will be sorted out and reborn totally independent of the others. The situation can be regarded as slightly similar to that of the human ear transmitting the sound from a whole orchestra in spite of the fact that its membrane can exist at only one place at a time.

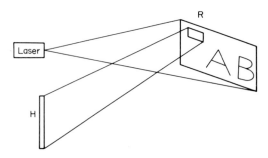

Fig. 2.4. The laser illuminates a screen with the letters AB and a mirror (R). The hologram plate (H) will be illuminated by both the reference beam (direct light from the mirror) and diffuse light from the letters.

2.2 Holograms of two-dimensional objects

It is now time to study holograms recording complicated objects instead of isolated laser beams. However, let us start by studying a hologram produced by utilizing the interference effect of collimated light beams. Let the object be a few white-painted letters bonded to a dark screen on to which a mirror is also attached (Fig. 2.4). Light will arrive at the holographic plate in the form of diffuse, scattered light from the letters (the object beam) and also in the form of well ordered light reflected from the mirror (the reference beam).

If the letters and the mirror are at a long distance from the hologram plate, then the light from the mirror and from each point of the object could be approximated as if it had arrived on the plate in the form of parallel beams. In that case each object point will, together with the reference beam, produce a diffraction grating on the hologram plate consisting of equidistant parallel interference fringes just like the gratings studied above.

After the plate is developed and fixed it will, if illuminated once again by the light from the mirror, diffract light in such directions that, to an observer behind the plate, it appears to come from the letters. If the letters are taken away or covered by black cloth, their holographic image will still be there. What is more important is that any beam of parallel laser light will reconstruct the image of the letters; it is by no means necessary that it is reflected by the mirror. Thus, by combining the object beam with an easily reproducible reference beam, we can reconstruct the image by once again illuminating the plate with a replica of the reference beam.

The collimated laser beam (a beam of parallel light) might well be the beam that is easiest to reproduce, and therefore a hologram made with collimated reference and object beams might represent the simplest of all holograms. This type of hologram is usually referred to as a Fourier hologram. The reason for this name is that at infinite distance any light distribution will

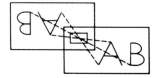

Fig. 2.5. The interference fringes recorded on the hologram plate will, when illuminated by direct laser light, reconstruct two images of the letters: the true virtual image and the conjugate image. Lines drawn from corresponding points on the two images will be divided in half by the image of the reference source. Compare with Fig. 2.1(c).

separate into its Fourier components, e.g. the sharp shadow of a knife-edge will be transformed into a diffraction pattern consisting of a set of interference fringes.

2.2.1 True and false images

The view through the processed plate towards the laser shows the letters A and B of Fig. 2.4 at the side of the laser beam. This image represents the true reconstruction of the object beam. On the opposite side of the origin of the reconstruction beam is seen another set of letters, as shown in Fig. 2.5. This image, the conjugate image, represents the false beam described earlier, which could have been excluded if a sufficiently thick emulsion had been used. If the hologram plate is turned around so that the emulsion is towards the observer, it is still possible to see exactly the same pair of images, but mirror reversed. All these images will be virtual and appear to exist at infinite distance as seen through the hologram plate.

There is another way to reconstruct and study the images and that is by shining the undiverged laser beam through the plate and looking at a screen behind. On the screen will be seen the bright spot from the direct laser beam and on each side of it the two pairs of letters. In this case the image appears to be a real image that can be recorded directly on ordinary photographic film or paper. The reason is that the small illuminated area of the hologram plate works like a small aperture which, as in a camera in which the lens has been exchanged for a pinhole, focuses the image at any distance. If, instead, a large area of the plate is illuminated by a collimated beam of large diameter, then the image will be blurred at short distances. In that case the true image would be sharp only if the screen was placed at the same distance from the plate as the letters had been during the exposure of the hologram plate. The reconstruction beam should have a direction that is antiparallel to that of the reference beam.

Let the direct laser beam be reflected by the emulsion side of the hologram plate and then directed on to a screen. In this case the two images of the pairs of letters will also be seen projected at each side of the laser beam.

In all cases the relation between the two images will be simple. From every point on the true image a line can be drawn through the bright spot of the reconstruction beam to the corresponding point on the false image. These points on the two images will always be at identical distances from the reconstruction beam.

The hologram under discussion was made by placing the objects and the source of the reference beam at such large distances, compared to the size of the hologram plate, that the two beams appear to be collimated and all the wavefronts to be flat. This type of hologram is usually referred to as a Fourier hologram because at large distances the light is separated into its Fourier components. A significant characteristic of this type of hologram is that the image is only two-dimensional; it does not possess any three-dimensional characteristics. The image appears to be at infinite distance, i.e. there is no parallax, because the information is identical over the whole plate.

Positioning the object at an infinite distance during exposure certainly involves practical problems and therefore one usually places the object in the focal plane of a positive lens or mirror. The result will be that every point on the opposite side of the lens will give rise to a beam of parallel rays in which the hologram plate is located. Different points on the object produce beams at different angles. Thus it is said that the object is placed in the focal plane of the lens while the hologram plate is positioned in the Fourier plane of the object. Usually the plate is placed at the focal point opposite to that of the object. Such a configuration is, of course, not possible for three-dimensional objects but can only be used if they are two-dimensional.

2.2.2 Information content of thin Fourier holograms

Fourier holograms are often thought of as carriers of information in optical memories and therefore it might be interesting to calculate how many beams

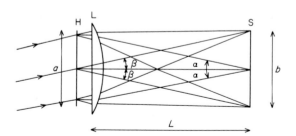

Fig. 2.6. The information capacity of a thin Fourier hologram is studied. A hologram plate (H) in front of a lens (L) produces diffraction-limited points on a screen (S). It is calculated that the required number of points on the hologram plate is of the same order of magnitude as those that are to be resolved on the screen. Thus, in this case, holography could carry no more information per unit area than photography.

a certain area of a hologram can produce. Let a thin Fourier hologram with area $a \times a$ cm^2 be placed in contact with a lens (Fig. 2.6) that, using diffraction-limited resolution (Section 1.3.3), focuses the beam on to a screen with area $b \times b$ cm^2. The distance separating the hologram from the screen is L cm. The closest fringe spacing (d_s) that the lens can produce at the screen is that formed by two intersecting beams coming from points separated by the lens diameter a. Thus:

$$d_s = \frac{0.5\lambda}{\sin \alpha} \simeq \frac{\lambda L}{a}. \tag{2.4}$$

The number of fringes on the screen will be n_s:

$$n_s = \frac{ba}{\lambda L}. \tag{2.5}$$

The separation of fringes (d_h) needed at the hologram plate to produce a deflection over half the screen is

$$d_h = \frac{0.5\lambda}{\sin \beta} \simeq \frac{\lambda L}{b}. \tag{2.6}$$

The number of fringes (n_h) on the holographic plate necessary to deflect the beam over the total screen will be

$$n_h = 2ab/\lambda L. \tag{2.7}$$

Thus it is found that the number of lines on the hologram plate (n_h) has to be twice the number of resolvable lines on the screen (n_s). If, instead, the three-dimensional case is studied, in which the lines are concentrated into points, it is found that the number of information points on the hologram plate has to be four times the number of resolvable points on any information receiver. This result is independent of the distance separating the hologram from the receiver and also of their relative sizes.

If hologram and receiver are of identical size it would thus be possible to store four times more information on the receiver directly, by covering it with a photographic plate, than by using holography. The information could then be recorded, for example, in the form of transparent points on a dark plate. If a certain amount of information in digital form is to be recorded by holography, then four times more information points are needed than when ordinary photography is used. The number of information points would have been equal for both methods if the holographic conjugate image did not have to be eliminated.

Thus it can be seen that there is no extreme information capacity hidden in the holographic process. What then is the advantage of holographic memories? Is it the fact that the information is not localized to any certain point on the hologram plate but spread uniformly over its total area? (In a

Fourier hologram the information is, as stated previously, identical over the whole plate.) The result of this is that local damage will not totally destroy any specific information. However, this statement is true only if the maximum information capacity is not utilized. If it is, any damage to the plate would lower the resolution and thus destroy information. On the other hand, if the total capacity is not used, but only half of it, then some local damage could be accepted. However, local damage could also be accepted in ordinary photographic recording, e.g. by repeating the information twice at different localizations on the plate.

Thus it is by no means self-evident that holography is superior to ordinary photographic recordings. A thick hologram, a volume hologram, could, however, record much more information than the thin hologram studied to date. This statement is true, but a volume of photographic material could also be supplied by the use of a number of photographic plates.

It is of course very difficult to arrive at a final conclusion as to whether there are any advantages in the use of holographic memories. Probably the answer lies in the practical possibilities of making the stored information available in a short time. There are good reasons to believe that the hologram's unique ability to reconstruct information when requested to do so by a certain input (e.g. angle and localization of a reconstruction beam) will be of the greatest importance.

A simplified sketch of a holographic memory is given in Fig. 2.7. The beam from a laser is deflected by an electronic device, e.g. a Bragg cell (B), towards an array of small holograms (H), each of which carries information in the form of bright dots that are projected on to a detector array D. The Bragg cell consists of a crystal coupled to an ultrasound generator. A standing wave pattern inside the crystal produced a pattern of alternating high and low refractivity which works like a grating or like the simple hologram of Fig. 2.6. Thus, the number of standing waves within the crystal has to be at least

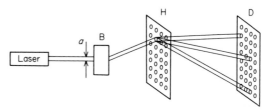

Fig. 2.7. A simplified sketch of a holographic memory. The laser shines a beam through a Bragg cell (B) which, controlled by electronic signals, deflects the beam to any one of a matrix of small holograms. Each hologram represents a memory that deflects the beam to any combination of detectors at D. In theory the response time of the device depends only on the number of small holograms and the frequency of the ultrasound controlling the Bragg cell.

double the number of requisite resolvable lines in the hologram array. Two Bragg cells coupled in series, one for x- and one for y-deflection, make it possible for the beam to reach any point at the hologram array. The response time of the device is limited by the time it takes for the ultrasound to pass across the crystal.

It is possible to calculate this response time (t). Assume that the number of resolvable lines required is N, the length of the utilized part of the Bragg cell is a, the speed of sound is v and finally that the frequency is f. The number of standing waves required will then be $2N$, while their separation will be $v/2f$. Thus

$$a = 2Nv/2f \qquad (2.8)$$

and

$$t = a/v = N/f. \qquad (2.9)$$

A Bragg cell can, however, for practical reasons, work only within a limited frequency range, and therefore f should be exchanged for Δf. Thus

$$t = N/\Delta f. \qquad (2.10)$$

Up to this point this book has concerned itself only with the special case in which the holograms record and reconstruct parallel beams. The next section will see an expansion to the general case in which there are no limitations on the shape of the light beams.

2.3 Recordings of three-dimensional objects

A hologram can be made in the manner illustrated in Fig. 2.8. At the left is a laser, the beam of which is made divergent by a lens (L) so that it becomes wide enough to illuminate the whole of the white-painted object (O) and also a mirror (M) placed close to it. A hologram plate (H) is placed so that it can record the required view of the object. The mirror is directed in such a way that it reflects the laser light on to that plate.

Thus every point on the plate will receive at least two light beams: one is the well ordered laser light beam directly reflected by the mirror (the reference beam), the other is the mixture of all the rays arriving from every point on the diffusely scattering object (the object beam). If the reference beam at all places is stronger than the object beam, every ray from the object will be individually recorded by the plate in the form of sets of interference fringes.

After exposure, the plate is developed, fixed and again illuminated by laser light, this time only by the beam from the mirror (the former reference beam is now used as a reconstruction beam). Even if the object is removed (or

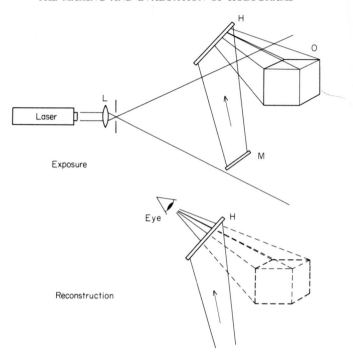

Fig. 2.8. The recording of a three-dimensional object. The beam of the laser is diverged by the lens (L) so that it can illuminate the whole object (O) and also a mirror (M) close to it. The hologram plate (H) is thus illuminated by both an easily reproducible beam from the mirror (the reference beam) and the complex diffuse light from the object (the object beam). These two beams produce a complicated interference pattern which is recorded by the plate. If the object is taken away and the processed plate is illuminated by a beam similar to that reflected by the mirror, the interference pattern on the plate diffracts light as if it came from the object so that an observing eye sees an image through it as if it has been "a window with a memory".

covered by a black cloth) an image will be seen at exactly the position where the object was during exposure. This image will be such a true replica of the real laser-illuminated object that it cannot be determined by eye or optical instruments that it is only an image! The three-dimensionality will be there *with its parallax*. By moving the head behind the plate it is possible to study the object from different angles. The recorded size and proportions of the object will be of such an extreme precision that measurements can be made with an accuracy in the order of fractions of a micrometre.

While a photograph represents just one view of an object, a hologram may represent thousands of views. The hologram is like a window (or mirror) with a memory. When we look through the hologram plate into the reconstructed world it is like looking into the real world as it was during exposure. Of course there *are* limitations. The hologram records the object as it looks illuminated only by the laser light. The object also appears to be covered by a pattern of tiny black spots, speckles. But that is also the case in the real world when objects are illuminated by the light from a laser.

2.3.1 Comparison to photography

Let us compare holography to photography (Fig. 2.9). The object, a cube, is seen in the middle. It is photographed from the right by an ordinary camera, while it is holographed from the left by a hologram plate which is also illuminated by a reference beam. The camera lens makes an image of the object on the photographic film. All the light reaching the film has to pass through the lens. No information will therefore be recorded on the film concerning what the object looks like as seen from any point outside the lens aperture, e.g. from B.

To explain the image-forming process straight lines are simply drawn from different points on the object through the centre of the lens to the correspond-ing points of the image on the film. Thus all the information recorded on the film represents the image as seen from one single point, the centre of the lens. Therefore no parallax and no three-dimensionality are recorded.

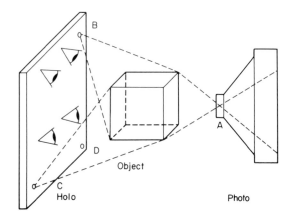

Fig. 2.9. Comparison between photography and holography. When a photograph is recorded by the camera at the right all the light has to pass through the lens. Thus straight lines can be drawn from every point on the object to every point where they are recorded on the film. These lines all pass through the centre of the lens (A). A two-dimensional image is formed because the only view recorded is that of the object as seen from A. The hologram at the left, on the other hand, has no information-limiting lens. At B the object is recorded as seen from above while at C and D it is recorded as seen from underneath. Thus a three-dimensional image is recorded at the plate and there exists no simple one-to-one relation between points on the object and points on the hologram plate.

In the holographic process, on the other hand, there is no lens between the object and the plate, as seen on the left of Fig. 2.9. Thus, the upper part of the plate (B) will record the object as seen from above, while a point at the lower part of the plate will "see" more of the underside of the object. Every point on the plate will record a slightly different view of the object.

Later, during reconstruction, when the plate is used to study the image the right eye will look through a part of the plate that recorded the object as seen slightly from the right, while the left eye will use the recording representing an image more from the left. Thus there will exist a parallax between the images seen by the two eyes and the object will appear three-dimensional, just like a stereoscopic image. But, while stereoscopy is based on only two images, holography represents an almost infinite number of images. If the observer's head is moved from left to right or up and down he will continuously see the reconstructed object from slightly different angles, just as if he were looking through a window. To study different depths in the image it is even necessary to focus the eyes at different distances, just as if the object really existed in space.

Now return to the conventional camera of Fig. 2.9. It can be seen that even if optical information is radiating from the object in all the directions of space, only a very small fraction will pass through the lens towards the film. However large the image on the film is made, the information content cannot be increased above the limit set by the lens aperture. The hologram, on the other hand, has no aperture that restricts the information radiating from the object to the plate. The larger the area of the hologram plate, the larger is the quantity of the information recorded. If the object were totally enclosed by holographic plates, the total content of the emitted optical information would be recorded. Thus, it is no wonder that the holographic process can utilize much more information than is possible by the use of ordinary photography.

2.3.2 Speckles: optical information quanta

It is often said that if a hologram plate is broken in pieces, each piece can be used to reconstruct the total image of the object. Let us take a critical closer look at this statement.

As shown in Fig. 2.9, there is a one-to-one relationship between the points on the photographic film and the points on the object. If, therefore, a small area on the film is destroyed, the corresponding area in the object scene is lost. In holograms, on the other hand, each point on the plate records the entire object. Thus a local destruction of the hologram plate does not destroy any particular point on the image. It is always possible to look around the defect on the plate as if it is a spot of dirt on a window.

Let us start to break a hologram into gradually smaller and smaller pieces while we study the reconstructed image. Using a large plate the total object scene is seen with full parallax; the object is seen from different directions as the observer's head is moved behind the plate. But when the hologram plate is broken into smaller pieces it will no longer permit large head motions before the eyes of the observer travel beyond its edges.

A still smaller piece of the hologram plate will restrict viewing to just one eye, thus the stereoscopic view is lost. It is, however, still possible to see some parallax as the eye is moved around behind the diminishing plate, so long as it is still larger than the pupil of the eye.

However, when the piece of the hologram plate becomes smaller than the pupil, then the resolution of the image starts to decrease, at the same time as the depth sensation caused by the focusing effect of the eye diminishes. In ordinary light a loss of resolution is seen in the form of a blurring of the image. In holography the image of the object appears to be built up by a random pattern of bright and dark irregular spots, speckles, and as the resolution decreases the number of speckles goes down. As the remaining piece of the broken hologram plate becomes smaller than the pupil of the eye, the speckles grow in size and finally a point is reached at which the object becomes as small as one speckle, in which case the information is completely lost.

The described effect is, however, not caused by the holographic process itself. It is rather caused by the fact that the object and the plate were illuminated by coherent light. Instead of looking through the hologram at the reconstructed image an observer could just as well look at the laser-illuminated object directly, or at its image reflected in a mirror. If this mirror is broken into successively smaller pieces the reflected three-dimensional image will be degraded in exactly the same way as described for the holographic image.

Using ordinary incoherent light the information capacity carried by the light rays has to be calculated. For the coherent light from a laser "optical information quanta" are carried in the form of speckles that can be studied and counted directly. A diffusely reflecting object that is illuminated by laser light radiates speckles in all directions. If the number of speckles per square centimetre is examined on a piece of paper illuminated by the object, this number will be found to increase with the square of the linear dimensions of the object and decrease with the square of the distance. A camera with a large lens that lets through a greater number of speckles will absorb a larger quantity of information and can therefore produce an image of higher resolution. In the following it will be proved that the number of speckles illuminating the lens area is identical to the number of speckles that build up the photographic image.

By studying the energy balance it was possible, earlier, to calculate the diffraction-limited resolution of a lens (Section 1.3.3). The reasoning was based on the assumption that no image detail could be sharper than the

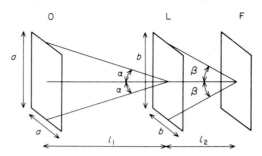

Fig. 2.10. The number of speckles projected by a scattering object surface (O) with area a^2 on to a lens L with area b^2 is calculated. It is shown that this number will be equal to the number of speckles seen at the image of the object focused by the lens on the screen. Thus, in this case, the number of speckles (optical information quanta) is conserved through the optical system.

separation of the interference fringes caused by laser beams radiating from two diametrically opposed points at the edges of the lens. This fact is just as true for the speckles: no speckle can be smaller than the closest possible fringes that are produced by light beams intersecting at the largest possible angle.

The number of speckles (n_L) projected by the scattering object (O) on to the lens (L) can be calculated by the use of Fig. 2.10. Referring to Section 1.1.4, the fringe separation d_L is given by

$$d_L = k^* \lambda/2 = \frac{\lambda}{2 \sin \alpha} \simeq \frac{\lambda l_1}{a} \tag{2.11}$$

and thus

$$n_L = \left(\frac{a}{d}\right)^2 = \left(\frac{ba}{\lambda l_1}\right)^2. \tag{2.12}$$

Let us now calculate the number of speckles (n_F) projected by the lens on to the image of the object (F). The smallest speckle is equal in diameter to the smallest possible separation between projected interference fringes, which is equal to d_F. But

$$d_F = k^* \lambda/2 = \frac{\lambda}{2 \sin \beta} \simeq \frac{\lambda l_2}{b}. \tag{2.13}$$

However, the size of the image is $a(l_1/l_2)$, and so

$$n_F = \left(\frac{al_2}{l_1} \frac{b}{\lambda l_2}\right)^2 = \left(\frac{ab}{\lambda l_1}\right)^2. \tag{2.14}$$

Thus $n_F = n_L$. This result proves that the number of speckles is conserved through the optical system and that the speckles really could be referred to as optical information quanta.

One must, however, remember that the statement is true only for the ideal case in which all the speckles illuminating the lens enter the camera and are used to build up the image. If there exist other laser-illuminated objects at such angles that they illuminate the lens but their light does not reach the film, then this light is lost and the corresponding number of speckles is scattered somewhere inside the camera.

The deduction proven above explains why the holographic process can record much more information about an object scene than is possible by the use of ordinary cameras. The reason is that it is much easier to make a hologram of a large area than a high quality lens of that same area, therefore many more information-carrying speckles will be recorded by the hologram than by the film.

2.3.3 Conjugate virtual image

Section 2.1 dealt with holography that used collimated beams and it was found to be possible to reconstruct two images from either side of the hologram plate plus two images by reflection at the emulsion side. All these images were focused at infinity. Now it will be shown what images can be reconstructed from a hologram produced by spherical waves and where these images are focused.

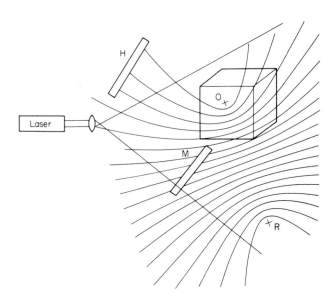

Fig. 2.11. The interference fringes recorded on the hologram plate (H) are caused by its intersection by a set of interference surfaces in the form of rotationally symmetric hyperboloids having one focal point (R) at the source of the reference beam (or its mirror image). The other focal point consists of the object point (O) under study.

First, there is the true virtual image that has already been described. This image is an extremely true replica of the real object with respect to form, size and distance. As described in later chapters, interferometric measurements prove that the errors can be essentially less than one wavelength of the laser light in use. This precision is, however, true only if the reconstruction beam is an exact copy of the reference beam as regards wavelength, divergence and angle of incidence on the hologram plate.

To see how this image is formed let us study one single point on the object, which will be the centre (O) of spherical waves that constitute the object beam (Fig. 2.11). The reference beam consists of spherical waves that are centred around point (R) from which the laser light is diverged, or from its mirror image as in the case of Fig. 2.8. Thus hyperboloids are formed in space, as seen in Fig. 2.11, having O and R as their foci. Where these hyperboloids intersect the plate, interference fringes are recorded. When these fringes are later illuminated from R they reconstruct the true virtual image of the object at O by diffraction.

Just as in the case with collimated beams, a false image will also be formed because the light is diffracted at identical angles to both sides of the reconstruction beam. It should, however, be kept in mind that these two diffraction angles are identical only if the reconstruction beam is normal to the plate. In Fig. 2.12(a) R is the point source of the reference beam while O_1 is a point on the object. H is the hologram plate, which is exposed and later placed back at its original position and reconstructed by a reconstruction beam from R; this reconstruction beam is identical to the former reference beam. The true virtual image of O_1 is observed to be exactly where the object was situated during exposure.

To find the false image, the conjugate image, it is assumed that the size of the hologram plate is unlimited. A normal to the plate is drawn through the reference source (R). The angle separating the object point from this normal is α. The conjugate image should also be found at an angle α, but on the other side of the reference source. This effect will be very similar to that described in Fig. 2.1, which showed holograms made by wavefronts from infinite distances. In Fig. 2.12(a), however, the limited distance to R and O_1 must also be taken into account in the following way.

A line is drawn through R and O_1 until it intersects the unlimited hologram plate of Fig. 2.12(a). This line will be the axis of the set of rotationally symmetric hyperboloids, formed by the two light sources. Where this axis intersects the plate the fringe frequency will be zero and there will be no diffraction effect. Consequently, the conjugate image also has to be situated on this same axis of rotational symmetry. The conjugate image of O_1, therefore, is found at O_1', where the axis intersects the line drawn at an angle α to the normal to the plate.

When the observer studies the two images through the plate he will see that the image at O_1 is an exact replica of the object, while the conjugate image at O'_1 will appear to be rotated, so that it is upside down, just as is seen in Fig. 2.5. However, the conjugate image will also appear to be larger, further away and inside out. The point on the object that was closest to R during

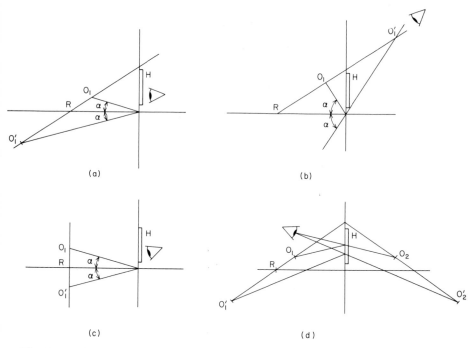

(a)

(b)

(c)

(d)

Fig. 2.12. (a) The conjugate image (O'_1) is found by drawing a straight line from the object point (O_1) under study through the point source of the reference beam (R). O'_1 will lie on this line. Draw the normal to the plate through R. The angle (α) between this normal and the direction to the object is measured. Draw a line from the foot of the normal at the angle $(-\alpha)$. O'_1 is found where this line intersects the line O_1R. The conjugate image will be virtual as it is on the same side of the plate as the illumination (R). See also Fig. 2.5.

(b) In this case the object (O_1) is closer to the plate (H) than in the previous figure. Using the same method it can be shown that the conjugate image (O'_1) falls on the obverse of the hologram plate. Thus the conjugate image will be real, and it could be directly projected on to a screen.

(c) If the object point (O_1) and the point source of the reference beam (R) are at identical distances then the conjugate image (O'_1) will be of the same size and appear at the same distance as the true image. This type of hologram is referred to as a "lensless Fourier hologram".

(d) When an ordinary transmission hologram is studied from the front side (the same side as the reference beam) a reflection image (O_2) is seen. It is accompanied by a conjugate image (O'_2) and both are positioned as if they were images of O_1 and O'_1 reflected by the hologram plate.

exposure will also be closest to R in the reconstructed conjugate image. Thus the depth impression is inverted: a convex surface will appear concave and vice versa. It is said that in this case the conjugate image is pseudoscopic.

If the definition of a real image is that it can be projected directly on to a screen, then neither one of the two images is real, because the hologram plate can have a focusing effect only on the light that has passed through it.

2.3.3 Conjugate real image

Figure 2.12(b) shows a holographic configuration that is slightly different from that of Fig. 2.12(a). The object O_1 is placed closer to the hologram plate during exposure. In this case the line through R and O will intersect the line representing diffraction in the negative α direction at the point O_1', which is at the observer's side of the unlimited hologram plate. Thus the conjugate image O_1' is a real image that can be projected directly on to a screen.

If the screen is taken away and the image studied directly, with the eye held close to the plate, then the image will appear to be turned upside down and the depth will be inverted just as is the virtual conjugate image of Fig. 2.12(a). As the eye is moved away from the plate, towards O_1', the image grows larger. At O_1' it is blurred out, and when the eye is a significant distance behind O_1' the image is seen again, but this time it is no longer turned upside down. The depth will, however, still be inverted.

If, during exposure, the source of the reference beam (R) and the object (O_1) are at the same distance from the plate, as seen in Fig. 2.12(c), then the true image and the conjugate image will appear at identical distances, the conjugate image being inverted and pseudoscopic. This type of hologram is referred to as a "lensless" Fourier hologram, because the angle separating object and reference beam will be approximately constant over the whole plate, just as is the case with a hologram made in the Fourier plane of a lens.

If, finally, the object is at the surface of the plate during exposure (e.g. focused on the plate by a lens), then the reconstructed true image and conjugate image will be identical and coincide to produce one single image focused on to the emulsion.

In the three-dimensional case there will also exist two images caused by reflection from the emulsion side of the plate, just as was described earlier for the two-dimensional images (Section 2.2.1). These images are caused by a combination of reflection and diffraction by the relief of the emulsion and will be seen at each side of the reflected reconstruction beam. They will be identical to the mirror images of the true and the conjugate images respectively (see Fig. 2.12(d)). It is the view of the author that these two mirror images $(O_2$ and $O_2')$ deserve much more interest than has so far been given to them in the literature.

The true virtual mirror image (O_2) coincides exactly with the real object as reflected by the emulsion side of the plate and is just as exact a replica of the object as is the ordinary diffracted true image. It has, however, the advantage that it can be studied without the necessity for light to pass through the hologram plate, which consequently does not need to be transparent. For the two mirror images to be seen clearly on ordinary hologram plates the angle between the reference beam and the object beam should be small, e.g. 20° or less. The true diffracted mirror image will coincide perfectly with the mirror image of the object if the reconstruction beam is identical to the reference beam during exposure.

2.3.5 True and false images explained by the use of the holo-diagram

It is now necessary to go back again and to study how all the images described in the preceding two sections can be explained using the hyperboloids and the ellipsoids of the holo-diagram. The ordinary true virtual image of Fig. 2.12(a) is caused by diffraction and reflection from the set of hyperboloids formed during exposure and having the point source of the reference beam (R) and the studied object point (O) as their focal points. The conjugate virtual image (O'_1) is formed at such a position that if the focal point of the hyperboloids had existed there it would have produced an almost identical pattern on the plate (H).

Thus it can be stated that the reconstruction beam cannot distinguish between those two patterns and accordingly reconstructs both images. It is clear that, to be similar, both sets of hyperboloids should be rotationally symmetric around the same axis, a line through the reference point and the object point. If the object point (O) is moved towards the reference point (R) along this axis, the fringe frequency at the plate will decrease and become zero as (O_1) and (R) coincide. When the object point has passed the reference point the frequency will again increase, and at a certain point (the conjugate image point O'_1) it will have the same frequency as during the original exposure. The fringes will also have an almost identical pattern because they will be in the form of rings centred close to the line $RO_1O'_1$, having zero frequency close to that centre.

The situation in Fig. 2.12(b) will be different. If the point O_1 is moved past R, the frequency at the hologram plate will never become as high as during the original exposure, not even if O_1 is moved to an infinite distance so that the hyperboloids are transformed into paraboloids with R as the focal point. If O_1 is imagined as passing through infinity, it will come back from the opposite side of the line RO_1, in which case the paraboloids have been transformed into a set of ellipsoids. At a certain position, the intersections of the plate by these ellipsoids produce a fringe pattern that is almost identical to the original pattern, and at that point the conjugate image O'_1 is formed. Thus, the conjugate real image O'_1 is formed because the reconstruction beam cannot distinguish the pattern on the plate caused by intersections by the hyperboloids from that caused by a certain set of ellipsoids. Accordingly, it produces a real image at the focal point of the ellipsoids.

Does there exist any set of fringe surfaces that are not rotationally symmetric around RO_1 but still produce intersections at the plate in the form of rings centred close to that line? Yes, the rings could just as well be centred around the point where the plate is tangential to the curved interference surfaces. This is exactly the case with the two mirror images. The reference point (R) of Fig. 2.12(d) is one of the foci of the set of hyperboloids while O_2 is the other. The unlimited hologram plate (H) will be tangential to these hyperboloids close to the extended line RO_1, and the intersections of the hyperboloids will be identical to those originally recorded on the plate (having R and O_2 as focal points).

Taking a closer look at this situation it can be proved that identical interference patterns are formed if one focal point of the hyperboloids is replaced by its mirror image as reflected by the plate. Here it is necessary to go back to the fundamental definition of the hyperbolas as being formed by the moiré effect of two sets of circles, each centred at one focal point.

Figure 2.13(a) illustrates how one of these circles is intersected by the hologram plate. If Fig. 2.13(a) is rotated through 180° about the plate HH, the result will be as seen in Fig. 2.13(b), where the centre of the circles is replaced by the mirror image of its position in Fig. 2.13(a). Thus the mirror image of the object point produces wavefronts that intersect the plate at exactly the same places as does the original object point. If these wavefronts interfere with another set of spherical wavefronts that are unchanged, then the moiré pattern and the interference pattern at the plate will also remain unchanged when the object point is interchanged with its mirror image.

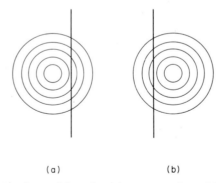

(a)   (b)

Fig. 2.13. The object point and its mirror image as reflected by the hologram plate produce the same intersections between this plate and the spherical wavefronts emanating from the point. Thus any transmission hologram also produces a true image in reflection. Compare with Figs 2.3 and 2.12(c).

2.3.6 Images formed by antiparallel reconstruction

If a hologram plate is exposed with its emulsion side towards the object but is later turned through 180° about a vertical axis so that the emulsion faces the observer, it will produce the true and the conjugate images. However, the angle of illumination of the reconstruction beam has to be changed such that if it was originally from the left it now has to be from the right. The true image will be inverted whereas the conjugate image will be the right way up.

Now let the hologram plate be rotated through 180° about an axis normal to its surface so that it is turned upside down, keeping the emulsion towards the object. The result will be that the true image is replaced by its conjugate. If the two images are close together, the true image will appear unchanged but inverted. If, however, the angle between the two images is large, the true image will be seen to be highly distorted. By reconstructing the hologram from various different angles it is possible to find a number of distorted images; however, they are of no practical value and will not be discussed any further.

There do exist, however, some more images that are undistorted and of practical importance: those that are formed if the direction of the light rays in the reconstruction beam is changed by 180° as compared to the direction of the reference beam. Thus a divergent reference beam is exchanged for a convergent reconstruction beam and vice versa. The true image and the conjugate image will then be reconstructed at their proper positions, but the virtual image is transformed into a real image and the real image into a virtual one. Let us reconstruct the hologram of Fig. 2.12(b) in that way.

Light is focused by the lens (L) towards the reference point (R) of Fig. 2.14.

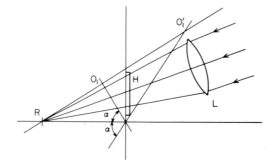

Fig. 2.14. The hologram of Fig. 2.12(b) is reconstructed by a beam that instead of diverging from R converges towards that point. The conjugate image O_1' is found in exactly the same way as in Fig. 2.12 but as the illumination is from the opposite side the real and the virtual images are interchanged. Light is focused by the lens (L) towards the true object image, which becomes real, while O_1' is the conjugate image, which becomes virtual. H is the hologram plate.

The true image (O_1), which was formerly a virtual image, will now be transformed into a real image that is projected exactly at the place of the original object. The conjugate real image (O_1') of Fig. 2.12(b) is transformed into a virtual image. The true real image will be an exact replica of the original object if the lens aberration and the influence of the glass plate of the hologram are neglected.

Suppose that a real image is required that can be reconstructed by a divergent beam. In that case use can be made of the equation given in Section 1.5.3:

$$(?) \quad \times \quad (?) \quad \times \quad (-1) \quad = (+1) \qquad (2.15)$$

| Object beam | Reference beam | Reconstructed beam | Real image |

image exposure

where the divergent beam is represented by -1, the convergent beam by $+1$, the virtual image by -1, and the real image by $+1$. Thus we see that the hologram could either be made in the ordinary way but using a convergent reference beam, i.e.

$$(-1) \times (+1) \times (-1) = (+1), \qquad (2.16)$$

or it could be made by using a convex lens to focus the object behind the plate (convergent object beam) using an ordinary divergent reference beam, i.e.

$$(+1) \times (-1) \times (-1) = (+1). \qquad (2.17)$$

When a virtual image is to be transformed into a real image it is often difficult to replace a divergent beam by a convergent beam of large area that focuses at a predetermined point with minimal error. Thus it might be advantageous to let the reference beam be collimated and simply change its direction by 180° before it is used as a reconstruction beam. If the formalism of the given equation is still required, the collimated light travelling in one direction should be referred to as $(+1)$ while the collimated light travelling in the opposite direction should be referred to as (-1).

2.3.7 Magnification

Up to this point we have restricted our studies almost exclusively to holograms that are reconstructed by a beam that is identical to the reference beam used during recording. One exception has been the case in which the direction of each light ray was made antiparallel so that a beam that diverges from a particular point is transformed into a beam that converges to that same point. Let us now examine what happens when the divergence and direction of the reconstruction beam are changed in a more general way or,

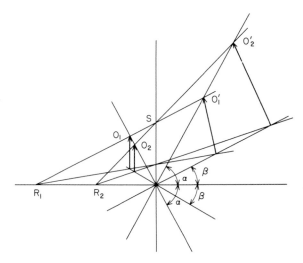

Fig. 2.15. The size, position and shape of the image O_1 are changed with the position of the reconstruction source R_1. The conjugate image O_1' is found using the method of Fig. 2.12. When the reconstruction source is moved from R_1 to R_2 the angle to the head of the arrow (α) and to its tail (β) will remain unchanged. The point on the hologram plate representing the axis of rotational symmetry (S) will also be unchanged. Thus the new true image O_2 and the new conjugate image O_2' are found by drawing a line from R_2 to S and studying its intersection with the line at the angle α.

in other words, what happens when the position of the source of the reconstruction beam is changed.

Figure 2.15 (which is similar to Fig. 2.12(b)) shows a holographic set-up that produces a true virtual image at O_1 and a conjugate real image at O_1' when the source of the reference beam is at R_1. If R_1 is moved towards R_2 the images will still exist along the line through the reference source (which is now R_2) and the point of zero frequency on the plate. Thus the virtual image O_1 is moved towards the plate to O_2, while the real image O_1' is moved away from the plate to O_2'. It will be apparent that a hologram behaves very much like a lens and that the following equation applies to the virtual image:

$$\frac{1}{r} - \frac{1}{o} = \frac{1}{f}, \tag{2.18}$$

where r is the distance between the plate and the source of the reference beam, o is the distance between plate and object, and f refers to the focal length of the plate which is determined during exposure and will remain constant independent of changes in the reconstruction configuration (it will, however, be influenced by wavelength changes). The only way to transform the virtual image into a real image is by letting r change sign and become smaller than f.

The real image obeys the following equation:

$$\frac{1}{r} + \frac{1}{o} = \frac{1}{f}.$$ (2.19)

If r is made smaller than f, the distance o between object and plate changes sign, so that the real image is transformed into a virtual image.

From Fig. 2.15 and from equations (2.18) and (2.19) it can be concluded that if the source of the reference beam is moved closer to the plate between the recording and the reconstruction, then the virtual image also will move towards the plate while it gets smaller. The real image, on the other hand, will move away from the plate at the same time as it is magnified.

A change in the wavelength of the reconstruction beam will also influence the size and the position of the images. If the wavelength of the reconstruction light beam is longer than that used during recording, the diffraction angles α_1 and α_2 of Fig. 2.15 will increase and thus the virtual image will grow larger and appear closer to the plate. The images reconstructed by a beam that is not identical to the reference beam will all be distorted; as a result there are practical limitations to magnification by changes in wavelength or beam divergence.

If the emulsion is thick, the efficiency of the holograms will fall off rapidly when the reconstruction angles are changed. The angle of each reconstructed ray is determined by the grating frequency, while the intensity depends on the angles of the recorded interference surfaces and on the degree to which the Bragg condition is fulfilled.

The thick hologram will, therefore, be properly reconstructed only by a perfect replica of the original reference beam or by a beam in which each ray moves in a direction that is antiparallel to that of the reference beam. There is, however, one more possible reconstruction of the thick hologram, and that is the case in which the reconstruction beam enters the emulsion at the same angle as did the object beam during recording. The result will be a highly distorted image reconstructed in the direction of the original reference source.

2.4 White light holograms

The different possible configurations that can be used to make a hologram can now be studied in a more general form. If the reference beam and the object beam consist of spherical waves, the situation will be as illustrated by Fig. 2.16, where hologram plates in two different positions are seen intersected by the hyperboloids. When the plate is in position A the configuration is similar to those used to make the holograms already discussed. This type of hologram is termed a transmission hologram,[1] because the reconstructed image is studied by the observer as if he was looking through a window.

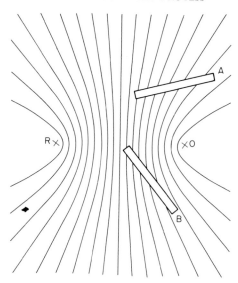

Fig. 2.16. A set of rotationally symmetric hyperboloids represent the fringe system caused by the interference of the reference and the object beams. One focal point is the point source of the reference beam (R), the other focal point is a point on the object (O). When the two points are at the same side of the plate an ordinary transmission hologram is recorded (A). If they are at opposite sides a Lippman–Denisyuk or reflection hologram is formed.

2.4.1 Lippmann–Denisyuk holograms

In the transmission hologram the two beams illuminate the emulsion from the same side and produce interference surfaces that are almost perpendicular to the plate. If, however, the hologram plate is placed in position B during recording (Fig. 2.16), then the two beams will enter the emulsion from opposite sides and produce hyperboloid interference surfaces almost parallel to the surface of the hologram plate. As described in Section 1.5.3, these surfaces will by reflection produce a virtual image of the object point when illuminated from the reference point. This type of hologram is called a reflection hologram, or a Lippmann–Denisyuk hologram.[2]

If the divergent reconstruction beam is replaced by a beam that converges towards R in Fig. 2.16, then a true real image will appear at O. The same rules apply as for the transmission holograms but the reflection hologram needs a thicker emulsion, to produce images with high efficiency. Therefore the conjugate image is so weak that it will have no practical influence, and the Lippmann–Denisyuk hologram produces just one image by reflection, either the true virtual image or the true real image.

When the reference beam is collimated, or divergent from a large distance,

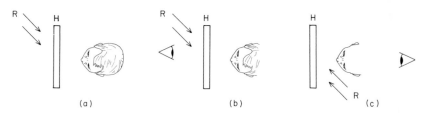

Fig. 2.17. (a) A reflection hologram of a person is recorded. The collimated reference beam (R) enters from the left of the hologram plate, while the person to be imaged is at the right. (b) When reconstructed by a beam (R) identical to the reference beam a true virtual image of the person is formed. (c) When reconstructed by a beam that is antiparallel to the reference beam the virtual image is transformed into a real one. If this image is examined from behind it should be possible to see the neck of the person imaged, but from the recording process (a) it can be understood that this is impossible. Only the front of the head was recorded and thus only the face can be seen, but from behind, as if it were a death mask. This type of image is referred to as pseudoscopic.

the distortion will be almost negligible provided that the real image is reconstructed by the same beam from the other side of the plate. One way to do this is simply by moving the source of illumination to the other side of the hologram plate, at the same time as the observer also moves to that side (see Fig. 2.17(c)). The depth of the image will, however, be inverted, so that the object will appear to be inside out. If the object is the face of a head, the image will look like a death mask, e.g. the nose being a depression instead of standing out. Let us have a closer look at this effect.

During exposure (Fig. 2.17(a)) of course only those parts of the head will be recorded that can be seen from the hologram plate. Thus the face, but not the back of the head, is recorded as illustrated in Fig. 2.17(b). When the image is reconstructed by a beam that is antiparallel to the reference beam a real image will appear positioned exactly where the head existed during exposure. However, as only the outer skin of the face was seen from the plate, this is the image observed, but seen as if "from the inside" (Fig. 2.17(c)).

These depth-inverted types of image are usually referred to as "pseudoscopic" in contrast to the "orthoscopic" images that represent our real world. What has been said here about pseudoscopic images reconstructed from Lippmann–Denisyuk holograms is, of course, just as true for the depth-inverted images of the transmission holograms described earlier.

One great advantage of Lippmann–Denisyuk holograms is that they can be reconstructed using white light. The reason for this fact is that they work by reflection from a great number of interference surfaces and that a high reflection efficiency is caused by the fact that these surfaces fulfil the Bragg conditions, as explained in Section 1.4.3. This means that the distances separating the interference surfaces are $0.5\lambda/\sin\alpha$, where λ is the wavelength of the light and α is both the angle of illumination and the angle of observation. Thus, for a point source of illumination the Bragg condition can be fulfilled for only one value of λ and therefore the image will be seen in only one colour—as if it had been reconstructed by monochromatic light. The Lippmann–Denisyuk hologram is therefore sometimes referred to as a "white light hologram" because it can be reconstructed by white light. The image is said to be achromatic.

2.4.2 Focused image hologram

The ordinary transmission hologram that works by diffraction produces an image that is a true replica of the object only if it is reconstructed by light having the same wavelength as that used during exposure.

If the wavelength of the reconstruction beam is longer, the diffracted angle α will be larger ($\alpha = \arcsin \lambda/2d$), so that the true virtual image of the object appears to be moved away from the source of the reference beam. It will also appear larger and closer to the plate.

If, however, an object is reconstructed using white light, it will be blurred because different colours produce images of different sizes and at different positions. The closer to the plate the object is located, the less influence a change of the diffraction angle will have during reconstruction. If the object, during exposure, is focused by a lens directly on to the emulsion, no change in λ could move an object point away from the position in the emulsion at which it is recorded.

Therefore a "focused image hologram" can be reconstructed in white light.[3] If, however, the object is three-dimensional, only parts of it can be focused sharply and thus details that reconstruct in front of, or behind, the plate will be blurred by colours. Thus focused image holograms can produce an approximately achromatic image in white light if the object does not have too large a depth, a few centimetres is usually acceptable.

One drawback with this type of hologram is that the lens from which the image is projected also represents the aperture through which the image can be seen (Section 2.3.1, Fig. 2.9). Thus the size of the lens sets a limit to the angle of observation. To produce an image with stereoscopic effect the lens should have a diameter that is at least larger than the separation of the eyes. A high quality lens of such size is quite expensive. One way to solve this problem is to replace the conventional lens by a holographic lens (Section 1.4.3) or to have no special lens at all but to use a two-step holographic method. First, an ordinary hologram (the master hologram) is made and a second hologram plate is exposed in its reconstructed real image plane. This second hologram will record a focused image hologram, the angle of observation being limited by the area of the master hologram through which the reconstruction is seen as through a window. The observer can walk up to this window to look through it, but he can never see the image from angles that are obscured by the window frame. The next section deals with this type of hologram.

2.4.3 Rainbow holograms

A relatively new type of hologram exists that also can be reconstructed in white light. These holograms are referred to as "rainbow holograms" because they usually produce an image the colour of which varies with the viewing angle.[4] The hologram is produced in the following way.

First, an ordinary transmission hologram (the master hologram) is recorded in the usual manner (Fig. 2.18(a)). After the plate has been processed the true real image is reconstructed by a beam that is antiparallel to the original reference beam (Fig. 2.18(b)). However, rather than the whole area of the hologram plate being illuminated by the reconstruction beam, only a horizontal slot about 10 mm wide is so illuminated. A second hologram plate is placed so that it intersects the real image, and a new recording is made on this plate, which in the usual way is also illuminated by a reference beam entering the emulsion from the same side as the rays producing the real image.

After this second hologram plate has been processed it is illuminated by white light that is antiparallel to the direction of the reference beam (Fig. 2.18(c)). This light will reconstruct the object, but only by those rays that are antiparallel to the object rays that illuminated the plate during recording. Thus light is diffracted only towards the real image of the thin horizontal slot of the master hologram. When white light is used for this reconstruction, many slots will be formed on top of each other in space, each one in its own colour. As red light is diffracted at the largest angle, in Fig. 2.18(c) the red slot will be seen on top of the other colours, while the violet slot will be furthest down.

When the observer places his eyes *in* the real image of the slot he sees the virtual image of the object reconstructed in only one colour, the colour that reaches his eyes. If he moves his eyes higher, the image colour shifts into the red. Observation *through* the image of the slot makes the virtual holographic image of the object almost monochromatic and there will be almost no blurring effect caused by different diffraction angles for different wavelengths.

Another advantage is the high reconstruction efficiency of this type of hologram, because most of the diffracted light is scattered in the direction of the eyes of the observer. One disadvantage is that vertical parallax has to be sacrificed: moving the head upwards will not result in the object being viewed as if from above. Horizontal parallax, and thus the stereoscopic view, will, however, be as usual as long as the observer does not try to observe from outside the real image of the slot.

Thus it could be said that in the interferoscope (Section 1.3.3) achromatic

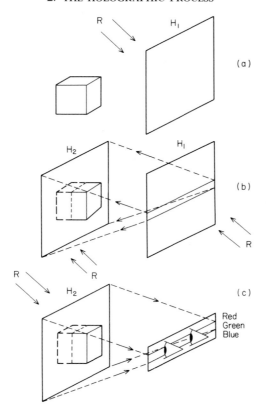

Fig. 2.18. (a) The recording of a rainbow hologram. A first hologram plate (H_1) is recorded in the usual way. The object is the cube at left while R represents the collimated reference beam. (b) A horizontal slot of the hologram (H_1) is illuminated by a reconstruction beam that is antiparallel to the reference beam of previous figure. Thus a real image of the cube is formed. A second hologram plate (H_2) is placed so that it intersects this image. It is illuminated by the reference beam R and a new recording is made. (c) The second hologram plate (H_2) is reconstructed by a beam that is antiparallel to the reference beam of Fig. 2.18(b). The image can be seen only when observed through the real image of the slot which is projected in front of the plate. One single colour is selected by the slot and thus the image can be seen even in white light. The reconstructions in (b) and (c) both produce pseudoscopic images and thus the net result is that a true image reappears.

interference fringes are seen in white light because of the dispersion (colour separation) of a prism. In the Lippmann–Denisyuk hologram the colours are separated by an interference filter effect, and one single colour is selected by the illumination angle. Finally, in the rainbow hologram, the colours are separated by diffraction and one single colour is selected by the observation angle.

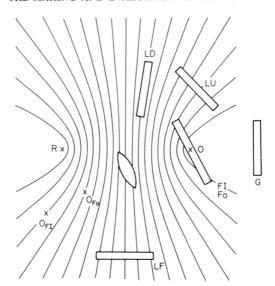

Fig. 2.19. Different holographic methods presented in the space covered by interference surfaces in the form of a set of hyperboloids, their common focal points being the point of the reference source R and a point on the object O. G refers to an in-line Gabor hologram, LU to a Leith–Upatnieks off-axis hologram, LF to a lensless Fourier hologram, FI to a focused image hologram, FO to a Fourier hologram, and LD to a Lippman–Denisyuk hologram.

2.4.4 Different holographic configurations

Let us finally sum up the different types of holographic configuration in one single figure (Fig. 2.19). As usual, O refers to an object point and R refers to the point source of the reference beam. O and R are the focal points of a set of hyperboloids representing interference surfaces in space, as described in Section 1.4.

The first type of hologram ever made was that produced by Dennis Gabor in 1948. This type of hologram (G) is usually referred to as a Gabor or in-line hologram because the reference beam is in line with the object, which has to be transparent. One disadvantage of this type of hologram is that the true and the conjugate images (in this case the virtual and the real images) coincide.

LU refers to the off-axis hologram discovered by Leith and Upatnieks. The conjugate image is moved away from the real image by twice the angle separating the reference beam and the object beam.

LF refers to a hologram that is called a lensless Fourier hologram because in many ways it behaves like a hologram made in the Fourier plane of a lens. The object point and the point source of the reference beam are at identical distances from the hologram plate.

FI is a focused image hologram (sometimes referred to simply as an image hologram) in which a lens focuses an image of the object (O_{FI}) at the hologram plate.

If instead the object (O_{Fo}) is placed at the focus of the lens, a Fourier hologram is formed (Fo).

The holograms produced in the vicinity of O and R are usually referred to as Fresnel holograms whereas those made at larger distances are referred to as Fraunhofer or Fourier holograms.

Ordinary holograms are sometimes called transmission holograms because the reconstruction beam passes through the plate when the image is studied.

When the plate is so positioned that object and reference beams enter from opposite sides the holograms are referred to as Lippmann–Denisyuk holograms (LD) or reflection holograms. In this case the reconstruction source is at the same side of the hologram as the observer and the hologram works more or less like a mirror during reconstruction.

References

1. E. Leith and J. Upatnieks. Reconstructed wavefronts and communication theory. *J. opt. Soc. Am.* **52**, 1123 (1962).
2. Y. Denisyuk. Photographic reconstruction of the optical properties of an object in its own scattered radiation field. *Sov. Phys.-Dokl.* **7**, 543 (1962).
3. L. Rosen. Focused-image holography with extended sources. *Appl. Phys. Lett.* **9**, 337 (1966).
4. S. Benton. Hologram reconstructions with extended light sources. *J. opt. Soc. Am.* **59**, 1545A (1969).

3

Holography for measurement of dimensions, deformations and vibrations

3.1 Conventional measurements on the holographic image

Before looking at the interferometric properties that are unique to holography
let us see how the holographic image can be used for dimensional measurements of the conventional type. As already mentioned, the holographic image
functions like a window with a memory and all the optical monochromatic
measurements that could be made through an ordinary window of equal size
can also be made through the hologram plate. If the hologram plate is of such
a high optical quality that the errors caused by aberrations in the glass can
be neglected, and if the reconstruction situation is identical to the exposure
situation (Section 2.3.1), then the precision of the measurements will be equal
to the diffraction-limited resolution (Section 1.1.1). When the object is
illuminated by laser light this limit will be of the same magnitude as the
speckle size (Section 2.3.2).

3.1.1 Focusing

A hologram plate of diameter D will permit an observation through a
telescope having a useful diameter of D and therefore the maximal resolution[1]
(r) will be

$$r = \frac{1.22\lambda}{2\sin\alpha} \sim \frac{1.22\lambda L}{D}, \tag{3.1}$$

where L is the distance between aperture and object. Exactly the same
resolution will be obtained if no telescope is used but the real image is formed

70

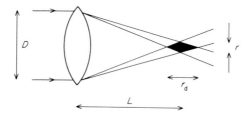

Fig. 3.1. The resolution of a lens with diameter D is calculated. The distance from the lens to the image is L and the two diagonals of a rhomb represent the transverse resolution (r) and the depth resolution (r_d) respectively.

by reconstruction with a beam that is antiparallel to the reference beam used during exposure and illuminates an area of the plate with diameter D.

The depth resolution r_d is reached if one point of the object is displaced along the axis of the lens so far that two diametrically opposed points on the periphery of the lens see it displaced by the distance r against a fixed background. From Fig. 3.1 it is found that

$$r_d/r = L/D \qquad (3.2)$$

and thus

$$r_d = 1.22\lambda L^2/D^2. \qquad (3.3)$$

This calculation gives the theoretical limit to the precision of depth measurements that are made by looking for the sharp image either when the virtual image is focused by a lens or when the real image is projected directly on to a screen.

Another practical method is to move a point-like object, e.g. a miniature lamp, around in the three-dimensional space of the virtual or the real holographic image and study at which coordinates it intersects a point on the reconstructed object. This point of intersection is determined either by focusing both eyes or by moving one eye from left to right behind the plate and looking for zero parallax. These methods are similar to those of ordinary stereography or photogrammetry but might be easier to use as no evaluation is necessary. The resolution will, however, be low because of the relatively small pupils of the eyes. The following section will examine the resolution limits in the space around two observing lenses.

3.1.2 Stereography and photogrammetry

Figure 3.2 illustrates how the depth resolution (r_d) can be calculated from the transverse resolution of two observing lenses. If the angular resolution of each lens is represented by sectors, the width of each sector will at any

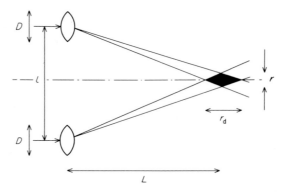

Fig. 3.2. The depth resolution (r_d) of stereography or photogrammetry is also represented by one diagonal of a rhomb, the other diagonal representing the transverse resolution (r) of each lens.

distance represent the transverse resolution. By studying equivalent triangles the following equations may be derived from Fig. 3.2:

$$\frac{0.5r_d}{0.5r} = \frac{L}{0.5l}. \tag{3.4}$$

Thus

$$r_d = 1.22\lambda L^2/0.5lD, \tag{3.5}$$

where r_d is the depth resolution, λ the wavelength, L the distance to the object from the lenses, D the lens diameter, and l the separation of the lenses.

If the two eyes of an observer are regarded as the centres of a set of resolution sectors, then the intersections of these sectors produce a set of

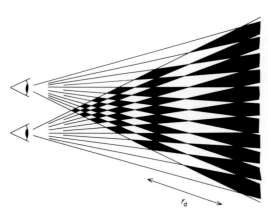

Fig. 3.3. The depth resolution of two observing eyes is represented by the moiré effect of two sets of radial fringes, each representing the transverse resolution of one eye.

rhombs one diagonal of which represents the transverse resolution while the other represents the depth resolution. In Fig. 3.3 every second sector has been printed in black so that the rhombs form a moiré pattern. The spacing of these moiré fringes represents the depth resolution.

In ordinary stereoscopy the angular view is limited by the construction of the eyes, but it is still interesting to study the resolution distribution in the space all around two observation points (A and B). The moiré pattern of Fig. 3.4 shows this distribution by visualizing the directions and magnitudes of both transverse and depth resolution. The depth resolution is represented by the separation of the moiré fringes (one diagonal of the rhombs) while the transverse resolution is represented by the other diagonal. It is obvious that the moiré fringes of Fig. 3.4 are analogous to the k-circles of the holo-diagram (Fig. 1.23), the reason is that the circumferential angle is constant along circles through A and B.

All the measuring methods described here are based on the assumption that an observer can distinguish details on the object surface. If the object has no contrast at all, the position of its surface cannot be determined by focusing, by stereoscopy or by photogrammetry. In all such cases its surface has to be covered by a resolvable pattern; for example, stripes or small dots could be painted or projected on to its surface. One example of such a measurement is a concave automobile part painted with small dark dots,

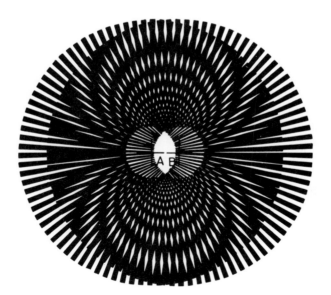

Fig. 3.4. The moiré pattern visualizes the distribution of direction and magnitude of both transverse and depth resolution in the space surrounding two observation points.

after which its holographic real image is reconstructed and projected directly on to a screen. This method permits the determination of a line of constant object depth in the form of an area of sharp focusing. (For other contouring methods see Section 7.4.)

3.1.3 Microscopy

Holography is sometimes used in combination with different optical systems, e.g. microscopes.[2] For example, when the living amoebae in a drop of water are to be studied through a microscope it can be advantageous to make a holographic recording.

If an ordinary camera is used for the study then at high magnifications, because of the short focusing depth, it is possible to achieve a sharp record only of a thin slice through the drop. If the whole drop is to be recorded it has to be photographed many times at different depths and therefore also at different points in time.

Using holography, on the other hand, the whole drop can be recorded at high magnification and at all depths. The drop can be illuminated by a single pulse from a ruby laser and the recorded image later studied at any depth by reconstruction. Because the hologram works like a window with a memory, the holographic plate can be positioned at any place at which it intersects the light beam that passes through the microscope. Thus the light rays are frozen into the plate in flight and during the later reconstruction these light rays are re-activated to continue their propagation through the microscope.

One way to make the hologram is to place the plate just over the drop so that it is recorded without any magnification. During reconstruction the microscope is focused through the plate towards the virtual image of the drop. One disadvantage of this set-up is that the whole magnification is used through a photographic plate with all its imperfections in the form of scratches and defects. Another disadvantage is that to gain a high magnification it is necessary to use an objective with a short focal length. Therefore it might become difficult to find sufficient space to position the plate between drop and objective, especially as the plate must also be illuminated by an unobstructed reference beam.

Another way to make the hologram is to place the plate at the other end, the ocular end, of the microscope. This means that the hologram would work almost like an ordinary photograph that records the magnified image. It would, however, have some properties of refocusing during observation, but this effect would be severely limited.

Finally, the hologram plate could be placed somewhere in between the objective and the ocular so long as it was not put at any image plane where the ocular would also magnify the plate defects sharply. In this way it is possible to record the whole drop holographically and later, by moving the ocular, to study the reconstruction at different depths using a high magnification. The resolution will be limited by the speckles, but this limit can be brought down to the same order of magnitude as the diffraction limit of the microscope objective.

3.2 Real-time hologram interferometry

It was described in Section 1.2.2 how a test glass could be used to measure the flatness of a metal surface. In that case the object surface was compared interferometrically to the surface of the test glass. Thus one surface was compared to another with which it had to be placed in *close contact*.

The Michelson interferometer (Section 1.3.4) brings the great advantage that the two surfaces that are to be compared interferometrically may be separated in space because one surface is compared to the *mirror image* of the other.

In hologram interferometry the object is either directly compared interferometrically with a *holographic image*, or two holographic images are compared to each other.

One great advantage of this procedure over other interferometric methods is that no complicated optics are needed. The hologram plate itself represents the beam-splitter, and the recorded image represents the surface to which the object is compared. Usually holograph interferometry is used to compare the object with itself at different points in time (time-lapse hologram interferometry) and in that case its surface may have any micro- and macroscopic topography.

In contrast to conventional interferometry there is no need for optically polished surfaces. Thus hologram interferometry makes it possible to measure differences in positions of, for example, a rough stone surface with an exactness of the order of a fraction of a micrometre. Even if the surface is so rough that its position cannot be defined with a resolution better than one millimetre, the hologram will remember all of its topography and react to any changes that are of the order of the wavelength of light or larger.

This result is remarkable, and still more impressive is that this high measuring resolution is achieved without touching the object, which can itself be many metres in size and standing a couple of metres away from the holographic apparatus.

The output from the holographic process is a three-dimensional image, covered by fringes that, like the contour lines of a map, reveal any displacements of the object surface. Thus one single hologram represents thousands of measuring points, all recorded simultaneously. From this image it is easy to find the points of extreme movement, and it is often this possibility which is of greatest value.

The holographic interferogram contains a large amount of information. A mere fraction of this information is usually enough to solve most measuring problems, e.g. to find defects, when holography is used in holographic non-destructive testing (HNDT). In other situations as much as possible is needed of the recorded information and relatively complicated evaluation methods are needed to reveal the magnitude and direction of the surface displacements in all three spatial directions. These methods will be discussed in the next section.

3.2.1 Measurement of displacement

If the deformation of an object is to be measured, it is very useful to start by first carrying out real-time hologram interferometry. Looking at the object through the processed singly exposed plate, interference fringes can be seen on the object directly as they are formed. Thus it is possible to study how large a force should be applied to produce such a number of fringes that they are easy to resolve, count and evaluate. It is also possible to study the stability of the set-up by knocking on the laboratory bench or by pressing with a finger on the object or the optical components. The fringes are caused by interference between the reconstructed image of the object and the object itself in real time.

Real-time holography[3] is slightly more difficult to use than double-exposure holography because a good plate-holder is needed to guarantee that the plate position will be identical during reconstruction and exposure. The procedure is as follows.

The plate is placed in the plate-holder, one exposure is made and the plate processed in the usual way. It is then placed back in the plate-holder and when the hologram is reconstructed by the reference beam used during exposure an image is seen in exactly the position of the real object. To see this image either the laser light is blocked off so that it does not reach the object, or the object is covered by a black cloth. If the laser light is then allowed to illuminate the object once again, it is seen to be covered by fringes caused by differences between the image and the real object. As the plate dries after the processing the fringes should disappear completely if there has been no change in the object and if the repositioning is perfect.

Usually one sees one or two residual fringes which can sometimes be eliminated by small changes of the plate position. Occasionally no fringes are seen at all because there was too large an error in repositioning. In that case the reconstruction should be examined to see if there are two images—or at least one image with a double contour. During adjustments the fringes should begin to form just as the double contour disappears. Start to look for the fringes at the reference mirror and its frame; it is here that they are usually seen first and at highest contrast.

Theoretically, if everything is perfect, the object should appear dark because at every place on the developed hologram plate where there is a bright primary fringe (the microscopic image-carrying fringe) there will be black silver grains. Thus the undeformed object should appear covered by one broad dark fringe, while a bright fringe should appear at all those object points that have been so displaced that the path length of the light has changed by half a wavelength or by an odd number of half-wavelengths.

However, shrinkage of the emulsion and other chemical processes make it uncertain what brightness should represent zero shift. This fact usually causes no problem as long as the fringe shift is constant over the whole image. The present author does *not* recommend *any* artificial method of drying the plate, e.g. by using alcohol or by using a hair-dryer, because that very easily causes emulsion changes that result in false fringes (for doubly exposed holograms such methods work quite well because then both recordings are affected to an identical degree).

There exist many ingenious techniques for simplification of the repositioning of the plate or even for elimination of this process completely. The present author recommends the type of plate-holder illustrated in Fig. 3.5. It is very stable because it has no slideways and no screws for micropositioning (see also Section 3.5.1). Alternatively, if the plate is to be developed *in situ*, it could be held in position by a clamp at its top and an open tank with developer could be lifted up until the plate is covered by the liquid. The same procedure is then repeated with the fixing bath.

Fig. 3.5. A hologram holder for real time and sandwich holography. The glass hologram plates are held in position solely by gravity and rest against three ball points and three cylindrical pins.

There are still more complex systems where the plate is immersed in a liquid gate (a cuvette filled with liquid) throughout exposure, processing and reconstruction. Commercial apparatuses are available that pump the different liquids in and out of the cuvette so that the development process is fully automated. To produce an image without false fringes it is important that during exposure and reconstruction the liquids have identical indices of refraction. In the opinion of the present author these methods of development and reconstruction of real-time holograms are needed only for very special cases, e.g. when the time between exposure and reconstruction must be very short, or if the process has to be automated, or if the accuracy of the measurement has to be within a small fraction of a fringe. In all other cases a simple robust plate-holder similar to the one seen in Fig. 3.5 is recommended.

3.2.2 Measurement of vibrations by stroboscopy and time averaging

Vibrations can be studied by real-time holography in at least two different ways. Let us start with the method that is slightly more difficult to use but easiest to understand because it is based on conventional stroboscopy.[4] The object is recorded in the ordinary way when it is stationary; the plate is processed; finally, the vibrating object is studied while it interferes with the reconstructed image. Interference fringes that reveal the displacement of every point on the object are of course formed all the time, but when the object vibrates the fringes move around at high velocities at the image of the object.

If the fringes are observed only during a short fraction of the vibration cycle then the fringes appear stationary. This result is achieved by the use of intermittent illumination, either by using a laser that can be pulsed at high rates or by interrupting the laser beam with a high speed shutter (a chopper). The brightness of these stroboscopic images will be lower than those of ordinary real-time holograms but the contrast of the fringes will be just as high. The sensitivity (number of fringes caused be a particular amplitude) will also be identical if the illumination pulses are synchronized in such a way that adjacent pulses arrive when the object is at each extreme point of displacement. By introducing a phase delay the sensitivity can be altered so that only a small part of a large amplitude is registered. A variation of the phase delay also makes it possible to study the phase relation between different vibrating areas on the object.

The following paragraphs describe how it is possible to study vibrations in real time even without stroboscopic illumination. This is a simpler technique but at the same time some advantages are lost, e.g. the possibility of revealing phase relationships.

To understand the *time-averaging* process let us start by first taking a closer look at the fringe-forming process when an object is deformed. To simplify

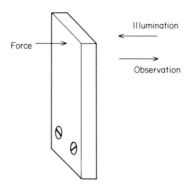

Fig. 3.6. A vertical steel bar is fixed at its lower end while illumination and observation are made in a direction perpendicular to its surface. Between the two exposures of the hologram a force is applied close to the top of the bar and the resulting interference fringes are studied.

the situation, consider the bending of a vertical steel bar that at its lower end is fixed by two screws to a rigid support (Fig. 3.6). Illumination and observation are carried out from such large distances that the ellipsoids of the holo-diagram are approximately flat (Fig. 3.7). Returning to Fig. 1.24(a), one

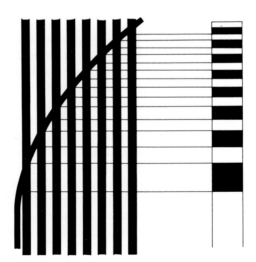

Fig. 3.7. To the left is seen how the bar of Fig. 3.6 is bent by the applied force so that it intersects imaginary interference surfaces. To the right are the resulting fringes. The interference surfaces to the left are parts of the ellipsoids of the holo-diagram; these are approximately flat and equidistant because of the large distances to the points of illumination and observation.

focal point (A) is the pinhole of the spatial filter while the other focal point (B) is the point of observation. Thus the steel bar appears to be surrounded by imaginary interference surfaces similar to those described in connection with ordinary interferometers (Section 1.1.4). These surfaces are flat, parallel and equidistant, their separation being 0.5λ. As the object is bent it will intersect these surfaces and each intersection causes one fringe in the real-time reconstruction. This fringe-forming process is illustrated in Fig. 3.7. In reality it is of course not necessary for the object to be flat when undeformed, but Fig. 3.7 is nevertheless useful to describe in an instructive way how the fringes behave as the deformation varies.

If the shape of the bar is identical to that which it had during exposure, then the image will appear to be fringe-free when it is studied through the processed plate. As an increased force is applied to the upper part of the bar it will alternatively become bright and dark. Each time this happens one bright fringe will move downwards, but it will never pass the fixed lower end of the bar. Thus an increasing number of fringes accumulates along the bar, like the pearls on a necklace. Their number is a measure of the displacement of the top of the bar while their location is a measure of the displacement at that point. By studying the variation in fringe spacing the bending radius of the bar can be calculated. Had it not been bent at all, but tilted instead, the fringe spacing would have been constant.

During the bending the downward motion of the fringes will be highest at the top and decrease asymptotically to zero at the lower end.

If the bar vibrates the fringes will move downwards during the forward motion, then they will stop and start moving upwards as the bending of the bar decreases again. The velocity will be highest as the bar passes through the undeformed state, then fringes start moving downwards again until the bar has reached its extreme backward deformation, after which the fringes move upwards again. If everything is symmetric there will be the same number, and the same position, of the fringes when the bar is at its extreme forward or its extreme backward position. Thus the fringes will appear stationary for a brief moment when the deflection of the bar is equal to the vibration amplitude.

When the oscillating fringe pattern is examined by eye or using a camera, if the exposure time covers a great number of oscillations, then only the stationary states will be registered. The situation will be rather similar to what happens if a pendulum clock is photographed using an exposure time that is many times longer than the time it takes for the pendulum to make one swing. The result will be a photograph of the clock having two more or less clear pendulums. During its swing the pendulum was blurred out because of its motion. Only at its turning positions was it recorded because there the velocity was zero.

Let us now return to our holographic vibration studies. The apparently stationary fringes are referred to as real-time time-averaged interference fringes. The vibration amplitude of the bar is found by counting the number of bright fringes and multiplying this value by 0.5λ. This statement is, however, an approximation. When the fringes approach their turning point they arrive *from* the top of the bar; as they leave they move *towards* the top. Thus they are registered as being displaced slightly towards the top of the bar or, in other words, each fringe represents slightly more than 0.5λ.

The fringes move with highest velocity closest to the top of the bar (at this point the number of fringes passing per cycle is at a maximum). Therefore the fringes representing the largest amplitudes will show the lowest contrast, while the contrast will be highest for those fringes that are closest to the lower stationary part. Thus the fringe contrast decreases with amplitude. The disadvantage of this phenomenon is that large amplitudes are difficult to measure. The advantage is that the non-vibrating parts (the nodes) are easy to find because they have the highest contrast. Ordinary double-exposure fringes (Section 3.3) are much more difficult to evaluate because usually the undisplaced parts are not disclosed by any difference in fringe contrast. The fringes representing a large displacement (higher-order fringes) usually have the same fringe contrast as the fringe at a stationary point (the zeroth-order fringe).

The technique of real-time time-averaged[5,6] holography is rather difficult for a beginner to use because the hologram has to be repositioned precisely — within one fringe. It is also necessary that the object vibrates symmetrically around its resting position during observation. If it is not possible to make the object appear fringe-free when the hologram is reconstructed after processing, the vibration modes can still be studied using the following method, but this does have a lower resolution.

Displace the hologram plate, or the object, or some optical component so that the non-vibrating object appears to be covered by closely spaced fringes. One simple way to do this is to introduce a tilted plane parallel glass plate in the reference or the object beam. When the object vibrates, all those parts that oscillate with an amplitude of zero or an integral number of values of 0.5λ will be covered by the closely spaced fringes. At other vibration amplitudes those fringes will be blurred out. The effect is rather difficult to see but at least the nodal lines can usually be found in this way.

Real-time hologram interferometry is most commonly used to find out how to arrange the holographic experiments. It is a very useful tool for the arrangement of holographic experiments and for deciding how to load an object during a static experiment. Real-time time-averaged holography can be used to find out at what frequencies there are resonant vibration nodes that could be of interest for further experiments. When these frequencies and the necessary driving forces have been found, ordinary time-averaged hologram interferometry (Section 3.4) can be used which does not depend on exact repositioning and which also produces fringes of higher quality.

3.2.3 Moiré pattern analogy

It is usually difficult to demonstrate the behaviour of interface fringes to a large audience and therefore the present author designed a moiré analogy experiment; the moiré pattern of Fig. 3.8 was photographed directly from a screen on to which it was projected by a View Graph during such an experiment. The apparatus consists of a Plexiglas plate covered by opaque

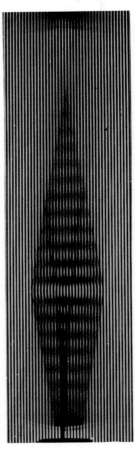

Fig. 3.8. The fringes caused by tilt are demonstrated with the use of the analogical moiré pattern apparatus. A Plexiglas rod covered by a grid of vertical fringes is tilted on top of a large sheet of Plexiglas covered by an identical grid. The moiré fringes are equidistant and the fringe contrast is constant over the bar.

Fig. 3.9 (right) The fringes caused by vibration are demonstrated by the analogical moiré pattern apparatus. The rod shown in Fig. 3.8 vibrates from left to right around an axis that is about one-third of its length from the base. The moiré fringes appear to be equidistant but in this case the fringe contrast decreases with the distance from the axis of vibration.

stripes spaced about 1 mm apart and is placed on the View Graph. On top of the plate is placed a Plexiglas rod that is also covered by an identical set of stripes. The Plexiglas rod is attached to a flat steel spring so that it can oscillate from left to right without touching the plate; its oscillation frequency is about 2 Hz. The stripes on the rod and on the plate are parallel and vertical when the object is not tilted.

When the rod is in its resting position no moiré fringes are visible, but when the rod is tilted more and more moiré fringes are formed at its top, from where they move down toward the rod's axis of rotation. The moiré fringes can, of course, not pass this axis, because at that point there is no motion perpendicular to the stripes (the primary fringes). Thus, by counting the fringes accumulated between the top of the rod and its axis of rotation, the displacement of the top of the rod can be found (Fig. 3.8). The contrast of all the moiré fringes (secondary fringes) is constant and independent of the fringe order.

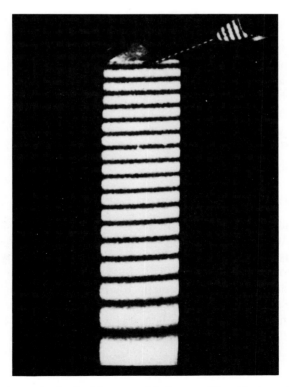

Fig. 3.10. Photograph of a doubly exposed hologram. The object was the bar of Figs 3.6 and 3.9. The fringes are not equidistant because the bar was not only tilted but also bent between the two exposures. The fringe contrast is constant over the object, as demonstrated by the moiré pattern analogy of Fig. 3.8.

The demonstration in Fig. 3.8 is equivalent to the real-time hologram interferometry of Fig. 3.7 in which the bar is illuminated and observed at large distances from the left. The stripes represent imaginary interference surfaces separated by 0.5λ. If illumination and observation had been made from shorter distances, the straight equally spaced stripes would have been exchanged for the ellipsoids of the holo-diagram (Fig. 1.24(a)).

If the rod had been tilted by hand from left to right, the moiré fringes would have moved up and down exactly as do the holographic fringes. When the oscillation is quick enough the moving fringes can no longer be seen because the reaction of the eyes is too slow. Only the stationary pattern representing the turning points is resolved. If the rod is now allowed to vibrate freely on its steel spring, the pattern shown in Fig. 3.9 is seen. The moiré fringes appear stationary and the contrast decreases with amplitude so that the zeroth fringe (the nodal line) has the highest contrast.

The fringe pattern is now an exact analogy to real-time time-averaged holographic interference fringes, and the relations between fringe order and amplitude can be calculated using identical equations. As the vibration amplitude slowly decreases the fringes grow broader while they move upwards and disappear from the top of the rod.

For a comparison of Fig. 3.8 and Fig. 3.9 with the real thing, consider the pattern seen in a hologram representing a steel bar that was slightly bent before the observation (Fig. 3.10) and an example of a time-averaged hologram, a turbine blade vibrated by the sound from a loudspeaker (Fig. 3.11). From Figs 3.10 and 3.11 it is easy to see the difference in fringe contrast between a static and a vibrating object and the analogy to Figs 3.8 and 3.9 is evident.

3.2.4 Comparison of different objects

If two different objects are to be compared by hologram interferometry, one essential condition is that the ordered macroscopic topography which is to be measured is not drowned in the interference pattern caused by the random microscopic topography of the object.

In ordinary hologram interferometry the two images interfere with each other because the hologram recognizes all the randomly distributed points on the object surface that scatters the light. If two objects are to be compared the hologram will no longer recognize any surface points and thus cannot identify any surface displacements. However, if the surface of the object is so smooth that it does not scatter any light but only reflects light in a mirror-like fashion, then two such objects can be compared to each other by hologram interferometry. They also could, in that case, be compared to each other in an interferometer that does not utilize holography at all. Whatever interferometric method is used, it is necessary to arrange the illumination such that

Fig. 3.11. Photograph of a time-averaged hologram. The object is a turbine blade set in vibration by a loudspeaker. The fringe contrast decreases with the distance from the fixed base in a similar fashion, as shown by the moiré pattern analogy of Fig. 3.9.

the specularly reflected light reaches the point of observation, because otherwise the non-scattering object will appear dark.

If an object is so rough that no macroscopic features can be seen on its surface when it is studied in an interferometer, then it cannot be compared to another object by hologram interferometry. The only remedy is to increase the separation (d) of the imaginary interference surfaces until only the required macroscopic topography is high enough to intersect them. From Section 1.1.4 it can be seen that

$$d = k\lambda/2 = \frac{\lambda}{2\cos\alpha}.$$

Thus d can be increased either by using a longer wavelength (λ) or a larger k-value, which means a larger angle (α) between the normal to the surface and the directions of illumination and observation respectively. Let us go back to the interferoscope described in Section 1.2.3 and study how this instrument can be used to determine what value of the angle α has to be used for holographic comparison of different objects.

Figure 1.12 shows a ground steel surface under study in the interferoscope. The object surface is compared interferometrically to a flat glass surface. By changing the angle of illumination and observation (α), the sensitivity (the distance d between the imaginary interference surfaces) is also changed. Figure 1.13 (in which $\alpha = 71.5°$, $k = 3.16$ and $d = 1\,\mu\text{m}$) reveals no macroscopic topography at all and thus a corresponding holographic recording could not be used for interferometric comparison between different objects.

In Fig. 1.13 (where $\alpha = 83.9°$, $k = 9.48$ and $d = 2\,\mu\text{m}$) the microstructure is beginning to disappear and in Fig. 1.14 (where $\alpha = 86.4°$, $k = 15.80$ and $d = 5\,\mu\text{m}$) the required macroscopic topography is clearly seen to be only

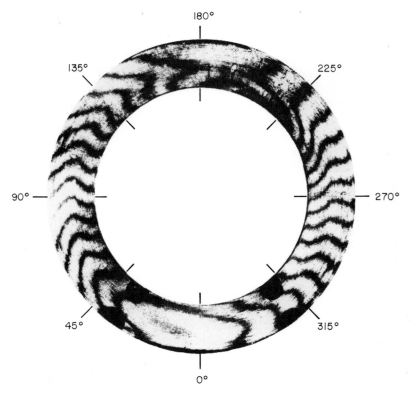

Fig. 3.12. By illuminating the inner side of cylinder bores with laser light at an oblique angle it is possible to compare one cylinder bore with another by holographic interferometry.[7] The k-value of the holo-diagram has to be sufficiently large, as demonstrated in Fig. 1.14.

slightly influenced by microscopic irregularities. Thus the angle of illumination and observation in Fig. 1.14 could well be used for holographic recordings in which the described object is compared interferometrically either to an ideally flat surface or to a similar object.

Finally, let us consider an early practical example of holographic comparison of different objects.[7] A cylinder bore that has been measured by other means and found to be within the requisite dimensional tolerances is used as a master to which other cylinder bores are to be compared holographically. The master is illuminated through a conical lens and the reflected light recorded by a hologram plate. The master is removed and the cylinder bore to be measured is placed in the same position by using a high precision fixture. Using real-time hologram interferometry an interference pattern is formed that reveals dimensional differences between the master and the object under study (see Fig. 3.12).

3.3 Double exposure

The hologram is exposed in a set-up similar to that shown in Fig. 2.7. After exposure of the hologram the object is deformed, e.g. by applying an internal pressure to the cube illustrated in Fig. 2.7, which could represent a pressure vessel. Without changing anything in the holographic set-up, a second exposure is made. Each exposure can be made by removing a shutter that hinders the light from passing from the laser to the diverging lens. After the two exposures the hologram plate (a glass plate covered on one side by photographic emulsion) is developed and fixed. When the image is reconstructed the object is seen covered by fringes that reveal the change in shape which it has undergone between the two exposures (Fig. 3.13).

During the first exposure the image-carrying fringes (primary fringes) are recorded on the hologram plate. The second exposure will add light to those fringes already recorded so that the effect is doubled. However, this statement is true only for those fringes that do not move in between the exposures. If one bright fringe moves by half a separation between earlier recorded fringes, so that it illuminates an area on the plate that was not illuminated before, then the two exposures will result in a uniform illumination of that area.

If, therefore, one point on the object has moved in such a way that the corresponding fringes on the plate move by half a fringe separation, then that point will be covered by a dark fringe during reconstruction because no fringes exist that diffract light towards that point. From this reasoning it can be understood that there is a moiré effect on the plate, between the image-carrying fringes from the two exposures.

The fringes on the plate will move by one fringe separation if the path length for the light transmitted from the laser to the hologram plate via the

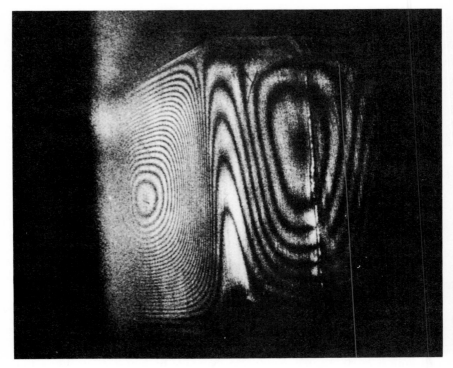

Fig. 3.13. A vessel with a volume of 10 litres. Between the two exposures 0.1 litre water was added to the original 9 litres, causing the vessel to expand. The depth of the holographic image is limited by the coherence length of the laser light.

object is changed by one wavelength. All parts of the object will appear bright during reconstruction at which that path length is unchanged or changed by a whole number of wavelengths. Thus all parts of the object that move between exposures such that they intersect zero or a whole number of the ellipsoids of the holo-diagram (Section 1.4.1) will appear bright. The parts corresponding to a change of path length of half a wavelength or an odd number of half-wavelengths will, on the other hand, appear dark during reconstruction. From this reasoning it can be understood that the interference pattern seen on the reconstructed image of the object is a moiré effect caused by the two sets of intersections between the stationary ellipsoids and the displaced object.

3.3.1 Holographic set-up

It is of great importance that the holographic set-up does not change between the two exposures so that no fringes are caused by factors other than the change(s) in the object. Let us study the stability demands if the tolerable variation in path length is set at 0.25λ, which corresponds to movement of the fringes by a factor of 0.25 times their spacing. The tolerance to changes during the exposure of high quality single-exposure holograms is of the same magnitude. If the primary fringes move by a factor of 0.5 times their separation they might get blurred out completely; the image then disappears. Even if the amount of movement is only half that distance again, the image becomes weak. Thus it is important that the optical components do not move during or between the exposures.

Consider the holographic set-up shown in Fig. 3.14 and the calculation of the resulting path length differences when different components are displaced. The beam from the laser at the left is divided by a beam-splitter (Bs) into the object beam, which passes straight through, and the reference beam, which is reflected. There are no high demands on the stability of the laser because any changes in its position will influence the phases of the reference beam and the object beam in the same way. The beam passing straight through the beam-splitter reaches a spatial filter (Sp_1) which consists of a positive lens that focuses the beam through a small pinhole, after which it diverges to illuminate the object (O). Any relative motion of the pinhole in relation to the object will result in movement of the image-carrying fringes at the hologram plate. Any motion of the beam-splitter or the mirrors M_1 and M_2 or the spatial filter in the reference beam (Sp_2) will also cause movement of the primary fringes.

The sensitivity to movement is highest for the mirror M_2, which reflects the

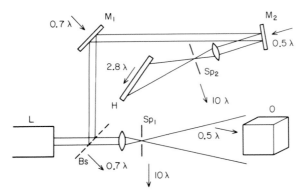

Fig. 3.14. A complex holographic set-up is studied for sensitivity to mechanical disturbances. The motion needed to produce a shift at one fringe is shown close to each component. The direction of maximum sensitivity is represented by an arrow. The beam from a laser (L) is split by the beam-splitter (Bs). The transmitted beam is diverged by the spatial filter (Sp_1), after which it illuminates the object (O). The beam reflected by Bs is also reflected by the two mirrors M_1 and M_2, after which it is diverged by the second spatial filter (Sp_2) to illuminate the hologram plate (H).

beam through almost 180°. A displacement of 0.5λ in this mirror will result in a fringe motion of one fringe separation. To produce the same result the beam-splitter and the mirror M_1 have to be displaced by 0.7λ. If the spatial filter Sp_1 produces such a divergence that the beam radius is 0.1 m at a distance of 1 m, then a sideways displacement of 10λ results in a movement of one fringe separation. The same result is valid for the spatial filter Sp_2. The directions of illumination and observation of the object (O) are almost parallel and the sensitivity of the object is such that a motion of 0.5λ (of the object) produces a fringe movement of one fringe separation.

Finally, the image-carrying fringes on the plate could also be destroyed by motions of the plate itself. The fringe spacing (d) depends on the largest angle (2α) between the rays of the object and the reference beams. Assuming 2α to be 20° in value, then

$$d = \frac{\lambda}{2 \sin \alpha} = 2.8\lambda.$$

The sensitivity to motions of the plate will be highest in the direction perpendicular to the primary fringes.

For each element Fig. 3.14 shows what direction and amplitude of displacement are needed to produce a path length difference of one wavelength at the hologram plate. It is evident that the risk of unwanted influence from unwanted movements grows with every optical element. The risk that differences in air temperature (temperature gradients) will have on the optical path lengths must also be taken into account. If the temperature is increased by 1 °C in the 1 m of air through which a He–Ne laser beam passes, the optical path length will decrease by about two wavelengths. Thus a difference in temperature of 0.5 °C between the reference and object beams (1 m in path

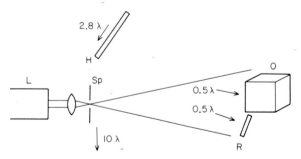

Fig. 3.15. This single-beam holographic set-up is recommended because of its simplicity and stability in comparison to that of Fig. 3.14. The beam from a laser (L) is diverged by a spatial filter (Sp), after which it illuminates the object (O) and a mirror (R) which reflects the reference beam towards the hologram plate (H). As in Fig. 3.14 the motion needed to produce a shift of one fringe is shown close to each component and the sensitivity direction is represented by an arrow.

length) would give rise to a fringe motion of one fringe separation. An eighth of this value or 0.06 °C is all that is acceptable in making a hologram of high quality.

Let us now leave Fig. 3.14 and instead consider the simpler Fig. 3.15, which is the type of holographic set-up that the present author uses almost exclusively when making holograms in difficult conditions. One beam-splitter, one mirror and one spatial filter are eliminated. If the angle α is kept small, the only optical element with a high degree of sensitivity is the mirror producing the reference beam (the reference mirror, R). This mirror is usually placed as close to the object as possible. Sometimes the reference mirror is even fixed to the object, in which case the sensitivity to translations of the object—and of the reference mirror—decreases to that of the hologram plate. In that case the total sensitivity to disturbances goes down to less than one tenth of that of the situation in Fig. 3.14. Finally, the simple arrangement of Fig. 3.15 has another important advantage. The reference beam passes through almost the same volume of air as does the object beam and thus the influence of differences in air temperature is minimized.

As has already been mentioned, clamping the reference mirror to the object simplifies the making of the hologram. One disadvantage, however, is that during reconstruction the image appears very close to the recontructing beam, and sometimes the conjugate image is seen covering the true image. In that case the set-up shown in Fig. 3.16 can provide a better solution. The reference mirror (R) is clamped to the centre of the object but directed to the side of the plate, where an extra mirror (M) reflects the reference beam towards the plate at a higher angle.

This method has the disadvantage that it adds the high sensitivity of the extra mirror. It also increases the sensitivity to movements of the plate. More importantly, however, the method retains a very low degree of sensitivity to

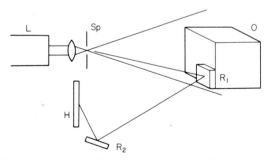

Fig. 3.16. Sensitivity to unwanted motions by the object is greatly reduced by clamping on to the object the mirror (R_1) that produces the reference beam. This beam can be reflected by a second mirror (R_2) so that when reaching the hologram plate (H) the angle separating object beam and reference beam is sufficiently large.

rigid body motion of the object and also to air temperature gradients. When the reference mirror moves together with the object the evaluation of the fringes becomes more complicated, but that problem can be solved by using sandwich holography, which will be described in Section 3.5.

When the two exposures of the hologram are of the same duration and the laser does not change its intensity, then the dark fringes should be totally black. This means that the image information is totally lost in those places. If that must be avoided, it is possible to let the two exposures be of different durations so that the contrast of the fringes is decreased and object details are seen even in the dark fringes.

If the double exposure is made using a pulsed ruby laser, then most of the problems about sensitivity to unwanted motions of the holographic set-up can be almost totally ignored. The pulse is usually of such a short duration ($c.\ 2 \times 10^{-8}$ s) that a hologram is recorded even if the object or the mirrors are moving about at a speed of some $4\,\mathrm{m\,s^{-1}}$!

If the two pulses are close together, e.g. having a separation of $c.$ 1 ms, then the demands for high stability are rather low; however, if the two pulses are separated by some seconds, then the need for stability is just as high as when a double exposure is made using a continuous laser. The reason is that in the latter case the limitations on the acceptable amount of movement are set by the fact that no fringes in the holographic image should be caused by any effects, between the two exposures, other than those motions of the object that are under study. There exist special methods to get around this problem and these make it possible to evaluate holograms even when rather large movements of optical equipment or of the whole object have occurred between the two exposures (see Section 6.5).

3.3.2 The use of two reference beams

When a hologram is doubly exposed both images are recorded by the emulsion in exactly the same way. Thus the direction of displacement, forwards or backwards, is lost because *the plate does not remember which exposure was made first.* Another drawback is that the two images cannot be separated. It is not possible to study just one of the images or to move one image in relation to the other. One way to get around this problem is to change the angle of the reference beam in between the two exposures.

Figure 3.17 provides a schematic representation of how to make such a doubly exposed hologram using two different reference beams. The first exposure is made using only the mirror R_1 to produce a reference beam. The beam from R_1 is cut out before the second exposure and instead a beam from the mirror R_2 is used. After processing the plate the object can be taken away and the virtual image is reconstructed by illuminating the hologram with the beam from mirror R_1. This image represents the object in its state during the

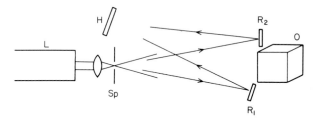

Fig. 3.17. The beam from the laser (L) reaches the spatial filter (Sp) which illuminates the object (O) and also the two mirrors R_1 and R_2 reflecting two reference beams towards the hologram plate (H). During the first exposure only R_1 is used; during the second only R_2. When the processed plate is illuminated by light from either only R_1 or only R_2 no fringes are seen; holographic interference fringes are formed if they are both used simultaneously.

first exposure. If, instead, the hologram is reconstructed only by the light from R_2, the object is seen in its state during the second exposure.

When the hologram is reconstructed by beams from R_1 and R_2 simultaneously then the object appears covered by fringes that reveal its changes in between the two exposures. By moving one of the mirrors with respect to the other it is possible to find the directions of motions of the object and to compensate for unwanted movements of the object or the holographic set-up. By slowly shifting the phase of one of the reconstruction beams, e.g. by moving one mirror continuously, it is possible to produce a continuous fringe motion in the image. This way phase-sensitive detectors capable of detecting a phase shift of $2\pi/100$ can be used to detect a change in a fringe position with an accuracy of a hundredth of a fringe.

It is, of course, much more practical if the mechanical translation of one of the mirrors in Fig. 3.17 can be eliminated so that the fringe motion can be continuous. One such solution is shown in Fig. 3.18, where D is a radial grating that deflects the light into two new beams. The zeroth-order beam (the beam passing straight through) is used to illuminate the object while the two diffracted beams each produce one reference beam at the hologram plate H. During exposure the radial grating is stationary but during reconstruction the zeroth-order beam is blocked off and the grating is rotated.

If the direction of the grating motion is as indicated by the arrow (V) in Fig. 3.18, the beam that is deflected upwards will be doppler shifted to a higher frequency while the frequency of the lower beam will be decreased. The doppler shifting of the two diffracted beams produces a beat signal when the reconstruction beams are recombined by diffraction at the hologram plate. The beat frequency can be calculated either from the doppler shift or by using the moiré pattern analogy (see Section 1.3.2). When the latter method is used it is easy to understand that the frequency of the beat signal will simply be

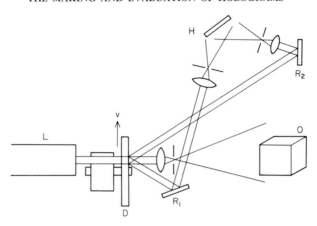

Fig. 3.18. A holographic set-up using the same principle as in Fig. 3.17. The beam from the laser (L) is split by a radial grating into two first-order diffracted beams which are used as reference beams which, after reflection by R_1 and R_2, illuminate the hologram plate (H). The beam passing through the grating illuminates the object (O). One of the reference beams is used at a time during the recordings whereas both are used simultaneously during reconstruction. When the radial grating is rotated the holographic interference fringes move. Phase-sensitive detectors make it possible to measure the fringe positions with extreme accuracy.

identical to the number of radial grooves that pass the laser beam in 1 s. Instead of the rotating radial grating a Bragg cell can be used as described at the end of Section 2.2.2. The frequency of the beat signal will in that case be identical to the number of sound waves passing per second, which, of course, is identical to the frequency of the ultrasound generator.

As was stated earlier, this "heterodyning" method[8] using two frequency-shifted reconstruction beams makes possible a measuring resolution of less than a hundredth of a fringe. However, in practice it is very difficult to utilize a resolution that is such a small fraction of a fringe. The mechanical stability has to be very high and, as a temperature difference between object and reference beams of 1 °C easily produces two fringes (see Section 3.3.1), it is obvious that the temperature gradients in the room have to be kept very low. In the present example the temperature difference should be less than 0.003 °C.

3.4 Time-averaged hologram interferometry

If an object is vibrating during one single, long exposure of a hologram, the reconstructed image will show fringes as if the object had existed simultaneously at approximately the two positions of maximal displacement which represent the stationary points during one vibration cycle.[9] The reason for this phenomenon is that the image-carrying fringes on the hologram plate, the primary fringes, are recorded only when they are stationary.

The situation will be rather similar to that of the pendulum clock already described in connection with real-time hologram interferometry. In that case it was the secondary fringes, representing deflection, that oscillated up and down and, like the photographed pendulum, were registered by eye or by camera only at their stationary points. However, when time-averaged holography is used directly it is the primary fringes that move and therefore are blurred out, and thus their images are recorded only at their stationary points. In consequence, two images are recorded in time-averaged holography and these two images—representing the vibrating object at its two extreme deflections—interfere with one another. The amplitude of vibration is defined as the distance between the neutral state and the maximal deflection. The interference fringes are caused by the peak-to-peak deflection and therefore each fringe represents an amplitude of 0.25λ.

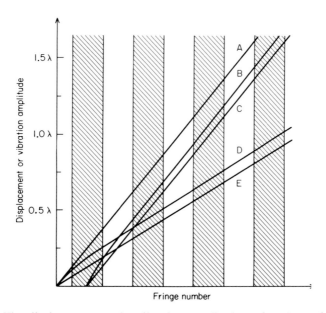

Fig. 3.19. The displacement or the vibration amplitude as functions of the fringe numbers. A is the double exposure displacement, B the real-time sinusoidal vibration amplitude, C the real-time displacement, D the time-averaged sinusoidal vibration amplitude, and E the time-averaged square wave vibration amplitude. Zero displacement produces darkness in real-time holography (B and C), whereas it produces brightness in all the other methods (A, D, E). The methods D and E are more sensitive than the others because their result is expressed in terms of amplitudes (half the peak-to-peak value) while they themselves react to total peak-to-peak value.

If the vibration was in the form of a square wave, then the above reasoning would be strictly true. Usually, however, mechanical vibrations are in the form of a sinusoidal wave, and in that case the above statement is an approximation. During the vibrations the primary fringes move from the position representing the neutral state to the position of maximal deflection and then they turn back again. Thus the fringes will be recorded just slightly displaced towards the neutral state, which means that each fringe represents an amplitude that is slightly more than 0.25λ. For practical purposes this small deviation is of any importance only for the first few fringes (see Fig. 3.19).

The interference fringes described above are found in the reconstructed image of a vibrated object which is continuously holographed during many oscillation cycles and are called time-averaged fringes; the method is known as time-averaged hologram interferometry. The contrast is highest for the zeroth-order fringe and decreases for the higher-order fringes. The reason is that the primary fringes, representing a high amplitude, will also move at a high velocity. Thus, the number of fringes passing for every cycle will be highest at the top of the bar shown in Fig. 3.7. The fringe contrast falls off more slowly than for real-time vibration fringes but faster than for fringes formed in a doubly exposed hologram of a stationary object.

Finally, the sensitivity of different methods such as real-time, real-time vibration, double-exposure, and time-averaged hologram interferometry are compared to each other in Fig. 3.19.

The moiré pattern analogy of Fig. 3.9 is, of course, not limited to real-time studies, but can also be used to demonstrate direct time-averaged holography. If the analogy is to be strictly true, however, the stationary Plexiglas plate with its striped pattern should be removed so that only the rod with its vertical stripes is left. When the rod is set into oscillation the stripes (the primary fringes) move to left and right. The primary fringes will be seen as stationary at those places where they stop at the same phase in the two stationary positions of the rod. At other places on the rod the primary fringes might stop at the left stationary point so that a dark strip covers exactly the position where a transparent strip had its stationary point when the rod was deflected to the right. At those places the stripes will be blurred out. Thus the rod will be seen to be covered by horizontal bands that are alternately fringe-free and covered by primary fringes, as seen in Fig. 3.20.

This situation is analogous to that of the image-carrying primary fringes in time-averaged holography, of which only those will be recorded that represent places on the object that vibrate such that the phase shift is zero or a whole number of wavelengths. To make the analogy complete, a transparency of Fig. 3.20 could be illuminated by laser light so that light will be diffracted by the stripes. If an observer places his eye in the first-order diffraction spot, only those parts of the bar will appear bright that are covered by fringes of high contrast. This method of observation (optical filtering) is analogous to the reconstruction stage of ordinary holograms, and the fringes on the bar will be analogous to time-averaged fringes.

Fig. 3.20. The moiré pattern analogy to time-averaged holography. A rod covered by a grid of vertical fringes is vibrated while being photographed using a long exposure. The fringes are re-solved only where the amplitude is such that they stay in identical positions at the two turning points. Thus the rod will be covered by areas that are alter-nately fringe-free or covered by fringes. By optical filtering (reconstruction) these bands are transformed into dark and bright fringes respectively.

3.5 Sandwich holography

Instead of making a double exposure on one hologram plate it is of course possible to make the two exposures on different plates.[10,11] By later combining the two reconstructed images it is then possible to study interference fringes caused by those dimensional differences of the object that have arisen between the two exposures. Such a method can be expected to bring a number of advantages.

It ought to be possible to find the direction of displacement, forward or backward, as long as it is remembered which plate represents the first exposure. In ordinary double exposure this information is lost because the plate does not remember which exposure was first. A memory of the time sequence could be recorded even in a double exposure if a known change (e.g. a large rigid body motion of the object, or a tilt of an optical component) is introduced between the two exposures, but such a method is bound to be rather complicated. In real-time holography the direction of displacement can be found by pressing the object with a finger and examining whether that extra force increases or decreases the unknown deflection. That method, however, only works as long as the holographic set-up is unchanged.

Another advantage is that a number of exposures can be made on different plates, each one recording the object under a particular load condition. Afterwards it becomes possible to study any combination of plates to measure how the object has changed in between just those two exposures. The object could, for example, be loaded by a force that is increased by one unit for each exposure. Afterwards the plates can be compared to each other in any combination, e.g. to study the deformation caused by the loading from 0 to 5 units, or from 5 to 10 units, or from 0 to 7 units, etc. Such a possibility does not exist directly using ordinary double exposures because then the load difference has to be decided before the second exposure. In real-time holography, on the other hand, it is possible to study many different load situations, but only as long as the holographic set-up remains undisturbed.

Perhaps the most important advantage of sandwich holography, however, is that when the exposures are recorded on different plates unwanted rigid body motions can be compensated for so that the fringes caused only by local deformations are left. The reason for this possibility lies in the fact that the two reconstructed images can be moved in relation to each other, and therefore a movement of the object between the exposures can be compensated for by a movement of one holographic image during reconstruction. This possibility for fringe manipulation corresponds to what has already been described with regard to ordinary interferometers (e.g. the Michelson interferometer of Section 1.2.4). The movement of one holographic image by changing the position of one plate is analogous in the interferometer to the movement of one reflected image by a change in the position of a mirror. In real-time holography it is also possible to manipulate the fringes, but that is more difficult to utilize in practice because of the holographic set-up, which, as already mentioned, has to be unchanged during the experiment.

The possibility of changing the interference fringe pattern during the reconstruction state brings still more advantages that will be described in later sections. There are, however, also the following disadvantages:

(1) The method is more complicated than that of double exposure because hologram plates have to be repositioned with high accuracy to guarantee a correct result.

(2) The built-in accuracy of a doubly exposed hologram that is untouched between exposures has, in the two-plate hologram, to be compensated for by recording reference points.

(3) In a double exposure the total mechanical and optical situation is identical for the two exposures, whereas change of plate can introduce errors caused by differences in plate thickness and plate inhomogeneities.

After testing different methods of combining the holographic images from different plates the present author found sandwich holography[12] to be most advantageous for practical measurements in industrial environments. The technique is relatively easy to use, the images are combined simply by placing one plate in front of the other so that their surfaces are in contact. Optical and mechanical errors are compensated for to a sufficiently high degree and the manipulation of the fringes is easy to perform without the use of any micro-positioning devices. Even if some errors can remain uncompensated for it should be kept in mind that an error in direction results in a measurement that is 200 % wrong, whereas the errors of the sandwich method are usually less than half a fringe. From Fig. 3.15 it can also be seen that the sensitivity of the hologram plate can be much less than that of the object. The following sections will describe how to expose and reconstruct sandwich holograms.

3.5.1 Exposures

Two hologram plates are placed in a plate-holder of the type already shown in Fig. 3.5, which is extremely stable and has no moving parts or adjustments. The hologram plate rests in position by gravity because even weak springs tend to bend the thin glass plate. It leans slightly (5°) backward against three supports in the form of short pins of hard metal with hemispherical ends. The plate is placed with its longer side resting on two 20 mm long hard metal pins with a diameter of 6 mm and positioned at different levels. Therefore, the plate slides down at an angle of 30° to the horizontal line until it comes to rest against a third pin. Eventually the plate rests by gravity on the three ball contacts at its back surface, on the two pins at its longer side, and on the single pin at its short side.

With this plate-holder it is easy to reposition hologram plates within one or two fringes for real-time holography or sandwich holography. Without the need for adjustments the same holder (or an identical one) is used to expose, reconstruct, and glue together the sandwich holograms. If a significant number of fringes are caused by repositioning, it is usually found that a small piece of glass has broken away at one of the points where the edge of the plate makes contact with one of the support pins.

The present author does not agree with those who find it difficult to make sandwich holograms because of repositioning problems and therefore have to invent alternative methods. Students are perfectly capable of using the plate-holder described, usually with the result that about eight times out of ten the errors will, without any adjustments, be less than two fringes. Even without any hologram holder, holding the plates by hand, with some patience it is possible to manipulate the two plates so that they form a hologram sandwich with no fringes on the reference surface.

When sandwich holography is attempted for the first time it can be useful to proceed as follows. First, one hologram plate is placed into the holder, its emulsion facing the object, its rear side resting against the spherical contact points, and two of its edges resting against the long support pins. Thereafter another plate is placed in front of the first, also with the emulsion facing the object. This second plate is placed on the long support pins and gently pushed with a finger at its centre until the surfaces of the two plates are pressed together, and the rear plate rests against the contact points. (Neither plate should have any antihalo coating.) It is important that the long support pins are of hard metal, otherwise they could be scratched by the glass edges, which would prevent them from sliding smoothly. After waiting a few minutes the first exposure is made.

The two plates are then taken away, and two new plates are placed in the holder in exactly the same way as the first two. After deformation of the object and a delay of a few minutes until everything has settled down, the second exposure is made. The four plates are subsequently developed and fixed in the conventional way.

3.5.2 Checking for repositioning errors

Now it is time to check the results. After the plates have dried (it is not advisable to use hot air or alcohol to speed the drying process of real-time or sandwich holograms), the first two plates are placed back in the plate-holder in the same way as prior to the first exposure. The object is blocked off, and the reference beam is used for reconstruction of the sandwich hologram. The holographic image of the object should be seen fringe-free or with most one fringe. If one of the plates is pushed sideways with a finger, away from one of the support pins, straight fringes should appear and disappear as the plate slides back again. If by mistake the two plates are interchanged before they are repositioned so that the plate which was exposed as a front plate (facing the object) is put behind the other during reconstruction, a number of dark concentric circles will appear covering the object image. Similar circles will also appear if the wrong (pseudoscopic) image is studied or if the plates are placed with their emulsions in the wrong direction. The two plates that were exposed after the deformation should pass through the same test programme.

If there is no fixed reference surface in the object scene, this examination of the fringes is essential to make sure that repositioning errors are not too large. When the combinations of the holograms described here result in no more than one or two fringes, there is good reason to believe that the sandwich hologram will be successful. However, if one of the two exposed pairs produces many fringes, something has gone wrong, and it is advisable to find out which plate is erroneous. For this purpose a holographic plate is required with the emulsion removed so that its glass base can be used as a substitute (compensation plate) for one plate of a sandwich hologram. This compensation plate is placed in the plate-holder. A front plate from one exposure is placed in front of it, and the real-time fringes[1] are studied. The reference surface, or undisplaced surfaces on the object, should be fringe-free or have at most one or two fringes. If the compensation plate has not been positioned behind the front plate, it will be repositioned too far from the object, which will therefore appear to be covered by fringes in the form of concentric circles.

The same procedure should be repeated with the rear plate of one exposure, which should be repositioned in the plate-holder behind the compensation plate. Once again there should be no, or only a few, real-time fringes. If the compensation plate is not used in front of the rear plate, a number of concentric interference circles will again appear. The reason for this is that the rear plate reconstructs the object as seen through the front plate during exposure. Therefore, the image of the object should, during real-time hologram interferometry, also be seen through a similar plate. These studies of real-time fringes are very useful to find which plate was placed in the wrong position and to learn how to reposition the plates correctly. Most errors are, of course, made prior to exposure, when the room has to be relatively dark.

After one or more of the checks described here have been made, it is time to combine the sandwich plates. The front plate from the second exposure should be placed in front of the back plate of the first exposure. After the object has been obscured, its image is studied using the reference beam for reconstruction. On the fixed reference surface and on undisplaced parts of the object, there should be no, or at most one or two fringes. These fringes in the reconstructed object space can now be manipulated by tilting the plate-holder.

If a permanent record of a particular sandwich combination is required the two plates can be bonded together using, for example, Cyanolite, which is a one-component cement based on cyanoacrylate. It is similar to Lock-tite and related to cements used for bonding strain gauges. When pressed into a thin film it hardens in less than 10 s. To bond the sandwich hologram together, the back plate is first placed in the holder, and two drops of the cement are applied close to two diagonal corners on the glass side of the front plate. This is then placed on the support pins and with a finger on the centre gently pushed toward and pressed against the rear plate in the plate-holder for a few seconds.

After the sandwich hologram is bonded it can be evaluated anywhere, without the use of a stable plate-holder. It is, however, important that it be illuminated by a reconstruction beam that has the same divergence as the reference beam had during exposure. This is accomplished most easily by arranging that the distance of the point from which the reconstruction beam diverges is the same as was the distance from the spatial filter to the hologram holder via the reference mirror during exposure. If these distances are not identical, interference rings and erroneous fringes will appear on the object. It is just as important that the wavelength of the reconstruction beam be identical to that used during exposure. If an argon laser is used for the exposure, reconstruction with a He–Ne laser will produce fringes that again have the form of concentric circles on undisturbed surfaces.

During the reconstruction of a sandwich hologram, the image from the emulsion of the front plate has to pass through the glass base before reaching the emulsion of the rear plate. Thus it is distorted by variations in glass thickness of the front plate. The rear plate, however, is reconstructed by a beam that has passed through the front plate, and therefore the two reconstructed images will have almost identical distortions when they interfere. The smaller the angle between object and reference beam the better this compensation works. Therefore the risk of erroneous fringes caused by glass defects is minimized if the object occupies a small angle of view and is placed close to the reference mirror. Thus we have found a fifth reason for a large distance between object and hologram holder.

When a good sandwich hologram is reconstructed correctly it should be possible to tilt it in such a way that not a single fringe is seen on the reference surface. Even if the eye is moved behind the hologram plate, no fringe should appear on that surface. Sometimes, however, one or two false fringes appear which do not make the use of the hologram impossible, but it is then particularly important that the eye or camera be kept close to the plate and in the same place during the observations. This test should be made prior to the evaluation, and it should be decided what part of the hologram to use. It should be pointed out that the observation must be made from a point close to the plates and that the starting point from which the sandwich tilt is measured is the angle at which no fringes are seen on the reference surface. The rules of tilt direction are true only if the first exposed rear plate is placed behind a second exposed front plate.

To save plates a hologram plate can be placed behind a compensation plate for the first exposure and another hologram plate in front of a compensation plate for the second exposure. One further advantage of this method is that the hologram image will improve slightly, because the rear plate does not have to "see" the object through the emulsion of the front plate. Still another advantage is that even antihalo-coated hologram plates can be used.

There is, however, one main disadvantage in this method: if many exposures are made with the hologram plates alternately in front of, or behind, the compensation plate, every combination cannot be used because the sandwich hologram must consist of one front and one rear plate.

3.5.3 Evaluation

Let us now study one example of how a sandwich hologram can be evaluated.[13–15] Figure 3.21 shows the object, which consists of one vertical,

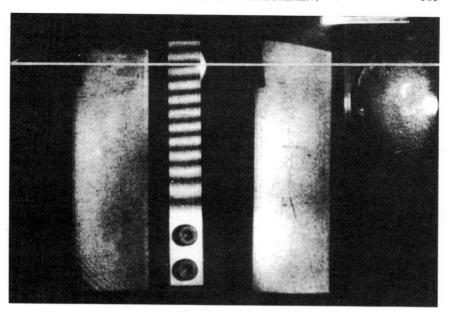

Fig. 3.21. Photograph of a reconstructed sandwich hologram. The test object in the middle was deformed 1 µm towards the observer prior to the exposure of the front plate. The rear plate (nearest the observer) represents zero deformation. The fringes should be counted from the base to the drawn horizontal line. Illumination and observation were parallel to the object displacement.

flat steel bar surrounded by two stable reference surfaces. The bar is fixed to the frame by two screws at its lower end. The information required is how the bar is bent and tilted in relation to the reference surfaces when a force is applied in its middle. The directions of illumination, observation and applied force are all perpendicular to the flat object. The distances to the illumination source and to the holograms are so large compared to the object size that the ellipsoids of the holo-diagram (Section 1.4.1) can be approximated into flat, parallel, equally spaced, imaginary interference surfaces separated by 0.5λ.

To understand the evaluation principle consider Fig. 3.22, in which the object (O) is in its resting position during the first exposure. One hologram plate (F_1) is placed in front of, and in contact with, another plate (B_1) in the plate-holder shown in Fig. 3.5. Both plates have their emulsions forward so that they are separated by the thickness (d) of the glass base of F_1. Let us now study the behaviour of one single speckle ray (Section 2.3.2) that is reflected horizontally from the object towards the hologram plates. Speckle rays behave as if they were reflected by small mirrors that are randomly distributed on the object surface. Thus a small tilt of the object will cause the rays to tilt by double that angle.

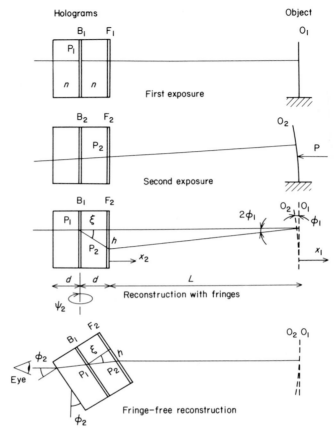

Fig. 3.22. The top of the object (O) is tilted at an angle ϕ_1 by the force P. Therefore a speckle ray from one object point is moved the vertical distance h from P_1 to P_2. B and F are the emulsions on the back and front plates, respectively. The glass plate thickness is d and the refractive index is n. ϕ_2 and ψ_2 represent sandwich rotation around a horizontal and a vertical axis respectively. Object translation is X_1. Corresponding sandwich translation is X_2. The identical reference and reconstruction beams are excluded from the figure.

After the first exposure has been made, during which the rear plate (B_1) is exposed through the transparent front plate, the two plates are taken away and replaced by two new plates F_2 and B_2. Prior to the second exposure the force is applied to the object, causing it to bend forward, its upper part being tilted through an angle ϕ_1. Thus the speckle ray under scrutiny is tilted downwards by $2\phi_1$. The result will be that the points where that same speckle ray intersects the two plates during the first and the second exposures is displaced downwards by a distance h.

All the plates are now processed and the plate F_2 is placed in front of the plate B_1. The plate-holder is in the same position as during exposure and the reference beam is used for reconstruction. When the sandwich hologram is reconstructed there will be two holographic images that interfere with one another because one image is deformed in relation to the other (see Fig. 3.22). This deformation is recorded on the hologram plates in the form of changes in the position of the speckles. By translating one plate in relation to the other it is therefore possible to change the fringe pattern representing the deformation.

This translation, or shearing, has to be made with micro-positioning devices because the sensitivity is almost interferometric (see Fig. 3.15). There is, however, a much simpler way to get that same precision: the two plates are bonded to each other and the whole sandwich hologram is tilted. This tilt will produce the same result as the shearing, but if the separation between the two emulsions is small, then the sensitivity will be low, so that the fringes can usually be manipulated while the plates are held by hand.

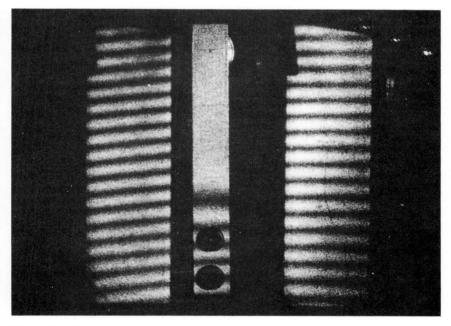

Fig. 3.23. The same hologram as that of Fig. 3.21 but the top of the sandwich hologram was tilted towards the observer during reconstruction. Fringes appeared on the reference objects but the number of fringes at the test object decreased. When the angle of tilt of the sandwich was ϕ_2 (Fig. 3.22) the fringes on the top of the object disappeared completely. Thus both the direction and the magnitude of the object deformation angle (ϕ_1 of Fig. 3.22) were revealed.

To measure the tilt of the top part of the steel bar (Fig. 3.21) the number of fringes (8) is counted between the point at which the force was applied and the horizontal line at its upper end. The angle ϕ_1 is found by dividing $8 \times 0.5\lambda$ by the distance separating the two horizontal lines crossing the bar. Another method is to tilt the bonded sandwich hologram until the top half of the bar is fringe-free and to measure the angle of tilt (ϕ_2) of the sandwich hologram at that state (see Fig. 3.22).

When the bar of Fig. 3.21 is fringe-free (Fig. 3.23) it is because at that angle the hologram reacts as if there had been no tilt of the object. The reason is that the sandwich is tilted in such a manner that the points at which that same speckle ray has intersected the two plates during the two exposures lie

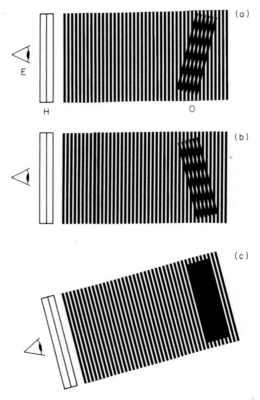

Fig. 3.24. The moiré pattern analogy to sandwich holography. If the object is tilted backwards (a) or forwards (b) between the two exposures, the eye (E) looking through the sandwich hologram will see the same fringe pattern during reconstruction. When the hologram is tilted, the imaginary interference surfaces tilt with it. If the object is tilted in a certain direction, there is only one sandwich tilt which produces no fringes, the direction that is analogous to that of the object.

along the same line of sight towards the object (Fig. 3.22). Using this simple assumption it is easy to calculate the relation between object tilt and sandwich tilt.

From Fig. 3.22 it is also seen that if the rear plate during reconstruction is the one that is first exposed, then the direction of object tilt and sandwich tilt for fringe-free reconstruction will simply be the same. The angle of tilt of the sandwich will usually be some 2000 times larger than the angle of tilt of the object, which makes it easy to measure.

Finally, let us calculate the relation between object tilt (ϕ_1) and sandwich tilt (ϕ_2) (see Fig. 3.22):

$$\tan 2\phi_1 = h/L \; ; \tag{3.6}$$

$$\tan \xi = h/d \; ; \tag{3.7}$$

$$n \sin \xi = \sin \phi_2 \; ; \tag{3.8}$$

where d is emulsion separation and L is the distance between hologram plates and the object. Therefore

$$\phi_1 = \frac{1}{2} \arctan \left\{ \frac{d}{L[(n/\sin \phi_2)^2 - 1]^{1/2}} \right\}, \tag{3.9}$$

For the calculation it has been assumed that the position of the reconstructed image is fixed during the sandwich tilt. This is, however, an approximation; the image moves with the sandwich rotation and therefore the measured ϕ_1 is slightly too large. This effect is very slight; it causes an error that is less than 1 %.

If ϕ_1 and ϕ_2 are both so small that the difference between ϕ and $\sin \phi$ can be neglected, then the following approximation can be used:

$$\phi_1 = (d/2Ln)\phi_2 . \tag{3.10}$$

3.5.4 Moiré pattern analogy

Finally, let us consider one more analogy to hologram interferometry. Figure 3.7 in an instructive way produces a correct analogy for the evaluation of interference fringes, but it does not give a true picture of the fringe-forming process. It is based on the approximation that the object surface during the first exposure is an ideal, flat, smooth surface. However, it has already been pointed out that one of the unique properties of hologram interferometry is that there is no need for such an ideal surface. Thus Fig. 3.24 will give an analogy that is closer to reality.

The object which might have a randomly curved surface is, as in Fig. 3.7, surrounded by flat, parallel, equally spaced, imaginary interference surfaces. The reason for these conditions is that illumination and observation are made from large distances. In Fig. 3.24 these surfaces are allowed to intersect the object and it is imagined that, during the first exposure, they become frozen within a coordinate system attached to the object. If, thereafter, the object is rotated or tilted, there will be a moiré effect between these frozen interference surfaces and the original imaginary fringe surfaces caused by the fixed points of illumination and observation.

This moiré effect, which is usually seen only on the outer surfaces of non-transparent objects, produces fringes that are analogous to the inter-ference fringes caused, for example, by tilting the object between the exposures of a doubly exposed hologram. Had the interference surfaces been the ellipsoids of the holo-diagram the analogy would be without any approxima-tions whatsoever. From Fig. 3.24(a) and (b), it is apparent that a small tilt of the object forwards or backwards will produce identical fringe patterns.

However, if the number of fringes could be set to zero, e.g. by tilting the imaginary interference surfaces in the same direction as the object was tilted, then it would be possible to find the direction of the tilt. There might exist two angles that produce a certain number of fringes, but there exists a unique angle that produces no fringes at all. This is the reason why the direction of rotation of an object can be found by sandwich holography. Figure 3.24(c) illustrates how a tilt of the sandwich hologram produces a tilt of the imaginary fringes. It also explains the simple analogy with respect to tilt direction: if the top of the object is tilted backwards, then the top of the sandwich hologram must also be tilted backwards to eliminate the fringes. However, the simple analogy of Fig. 3.24(c) does not show the magnifying system which results in the fact that the sandwich hologram has to be tilted at a much larger angle than the object, which, of course, is advantageous for practical evaluation.

References

1. M. Born and E. Wolf. *Principles of Optics.* Pergamon Press, New York (1965), p. 397.
2. R. van Ligten and H. Osterberg. Holographic microscopy. *Nature, Lond.* **211**, 282 (1966).
3. J. Burch. The application of lasers in production engineering. *Prod. Engng* **44**, 431 (1965).
4. E. Archbold and A. Ennos. Observation of surface vibration modes by stroboscopic hologram interferometry. *Nature, Lond.* **217**, 942 (1968).
5. K. Stetson and R. Powell. Interferometric hologram evaluation and real-time vibration of diffuse objects. *J. opt. Soc. Am.* **55**, 1694 (1965).
6. K. Biedermann and N. Molin. Combining hypersensitization and *in situ* processing for time-average observation in real-time hologram interferometry. *J. Phys. E, scient. Instrum.* **3**, 669 (1970).
7. E. Archbold, J. Burch and A. Ennos. Application of holography to the comparison of cylinder bores. *J. scient. Instrum.* **44**, 489 (1967).

8. B. Ballard. Double exposure holographic interferometry with separate reference beams. *J. appl. Phys.* **39**, 4846 (1968).

9. R. Dändliker, B. Ineichen and F. Mottier. High resolution hologram interferometry by electronic phase measurement. *Opt. Commun.* **9**, 412 (1973).

10. R. Powell and K. Stetson. Interferometric analysis by wavefront reconstruction. *J. opt. Soc. Am.* **55**, 1593 (1965).

11. J. Gates. Holographic phase recording by interference between reconstructed wavefronts from separate holograms, *Nature, Lond.* **220**, 473 (1968).

12. A. Havener and R. Radley. Dual hologram interferometry. *Opt. Electron.* **4**, 349 (1972).

13. N. Abramson. Sandwich hologram interferometry; a new dimension in holographic comparison. *Appl. Opt.* **13**, 2019 (1974).

14. N. Abramson. Sandwich hologram interferometry. 2. Some practical calculations. *Appl. Opt.* **14**, 981 (1975).

15. M. Dubas and W. Schumann. Contribution à l'étude théorique des images et des franges produites par deux hologrammes en sandwich. *Opt. Acta* **27**, 1193 (1977).

4

Evaluation of fringes caused by
out-of-plane motion

Holographic interference fringes are caused by phase shifts resulting from changes in path lengths. Thus the evaluation of the fringes caused by displacements is simply a question of the measurement and the calculation of changes in distances. There exist a great number of methods for finding the relationship between fringes and displacement, but some of these are very difficult to understand and to use because they involve many more factors than just the distances.

Scientific and engineering analyses are based on analogies. Mathematics is a very general analogy as it can be used in almost every field.

Because of its generality, however, the mathematical explanation sometimes becomes more complicated than the phenomenon which it is being used to explain. In those cases other analogies will be of great value as complements. Such tools as the holo-diagram, the string analogy and the moiré pattern analogy for fringe evaluation, and also the concept of wavefronts of observation and Young's fringes of observation will be or have been introduced to provide such complements. In the following different methods of fringe evaluation will be compared, enabling us to choose the one that appears most easy to use for each practical case.

The term "out-of-plane" motion is used because it has become accepted by holographers as referring to a component of motion that is perpendicular to the surface of the studied object, which is usually thought of as being at least approximately flat. The component that is parallel to the surface is referred to as the "in-plane motion". From the point of view of evaluation, however, it is more correct to speak about movements that are either parallel to, or perpendicular to, the line of sight. It is still better to refer to movements that are either parallel to, or perpendicular to, the ellipsoids of the holo-diagram.

In the following, if nothing is stated to the contrary, it will be assumed all the time that the holographic configuration is identical during recording and reconstruction, except for the changes in the object under study. In that case the hologram plate works like a window with a memory: during the reconstruction stage the light rays that were frozen during recording are "re-activated" and redispersed in exactly those directions which they had during exposure. Thus it is not necessary to include the plate and its position when the object displacement is calculated. The only influence of the plate is that its limited area restricts the number of possible angles of observation.

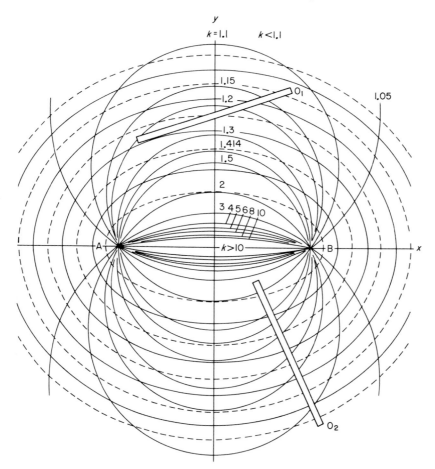

Fig. 4.1. The direction of increasing sensitivity of the holo-diagram is perpendicular to the ellipses of constant path length. Thus any opaque scattering object, e.g. O_1, is most sensitive to out-of-plane motions. To make it sensitive to in-plane motions it should be placed like O_2, which, however, makes observation impossible when A is the illuminating source and B the point of observation.

4.1 The fringe-forming process

Light travelling from the laser to the spatial filter is diverged so that it illuminates the total area to be studied on the object. Usually, the object scatters light in all directions and some of this light reaches the hologram plate, which it passes through (with or without a time lapse) to the eye of the observer or to a camera lens.

From the laser to the spatial filter all the rays travel by identical paths, therefore this distance has no influence on the calculation of variations of path lengths. Thus the holographic interference fringes are only caused by changes of path lengths for light travelling from the point of illumination (the pinhole of the spatial filter) to the point of observation via the object point under study.

Let us recapitulate regarding the holo-diagram described in Chapter 1 and study Fig. 4.1, in which A is the point source of illumination while B is the point of observation. The ellipses could be drawn by fixing one end of a string with one nail at the focal point A and the other at the focal point B. Keeping the string tight, and using a pencil, it is possible to draw the locus of constant string length from A to B via the pencil. Thus the path length from A to B is constant when travelled via any point along the ellipse.

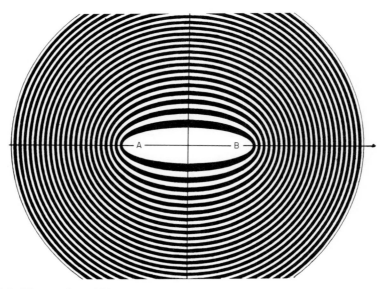

Fig. 4.2. The number of fringes seen between a fixed point and the displaced point of an object studied by holographic interferometry is simply a measure of the number of ellipses crossed by the displaced point. The flat equidistant interference surfaces of, for example, Figs 3.7 and 3.21 thus represent approximations which are possible because of large distances in combination with small objects.

Imagine that the string can be lengthened by just one wavelength and another ellipse can be drawn; this process can be repeated until a set of ellipses is formed around A and B. If every second area separating the ellipses is painted black the result will be that shown in Fig. 4.2. When an object point, e.g. in real-time hologram interferometry, is displaced so that it intersects n ellipses, then this number of fringes will also form between the displaced point and a fixed point on the object. The distance between adjacent ellipses varies with the directions to A and B, respectively. It is, however, constant along arcs of circles known as k-circles. If an object has n fringes between a displaced and a fixed point, then the displacement (d) in a direction perpendicular to the ellipses is given by

$$d = nk\lambda/2. \tag{4.1}$$

In reality, of course, the three-dimensional situation should be examined, in which case the ellipses are transformed into ellipsoids that are rotationally symmetric around the axis AB.

4.1.1 Scattering surfaces

If the surface of the object under examination was mirror-like, then it would reflect light from A to B only if it was curved in exactly the shape of one of the ellipsoids. If, on the other hand, the object was opaque and scattered the light diffusely in all directions, then it would direct some light towards B whatever its shape or position. However, the necessary conditions are, of course, that the surface under study should not be in a shadow and that it should be freely visible from B. Thus it is clear that a flat, opaque object could be recorded if it is placed almost parallel to the ellipsoids (O_1 in Fig. 4.1), but not if it is perpendicular to the ellipsoids (O_2).

It has already been mentioned that hologram interferometry has maximal sensitivity to movements of the object that are perpendicular to the ellipsoids, while the sensitivity parallel to the ellipsoids is zero. From this reasoning it is clear that if ordinary hologram interferometry is to be used to study objects that are opaque and diffusely scattering, then it is most useful for measurements of displacements that are perpendicular to the object surface (out-of-plane motion). In the following sections different methods for the evaluation of fringes caused by out-of-plane motion will be considered; the next chapter will treat methods for the measurement of in-plane motions which, however, all have a lower sensitivity.

4.1.2 The importance of defined path lengths

One necessary condition for interference fringes to be formed is that there is a path length difference and that this difference is well defined. If there is no

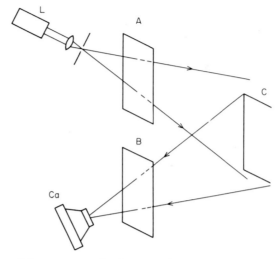

Fig. 4.3. The condition necessary for fringes to form is that the path lengths are well defined. Thus the light from the point of illumination (the spatial filter in front of the laser, L) to the point of observation (the camera, Ca) should pass through only one large scattering surface. In order to see the entire object it is, in the general case, also necessary that the light passes through one large scattering surface. Thus one and only one of the following areas should be scattering: the illumination surface (A), the object surface (C) or the surface of observation (B).

single definition of the path length, then there cannot exist any well defined fringes to evaluate. Let us study Fig. 4.3, in which A is a large scattering area of illumination, B is an area of observation, e.g. the hologram plate with its reference beam R, and C is the scattering object surface to be studied.

If A, B and C are all large, then the path length ACB cannot be defined. Every point on C is illuminated by rays having random phase distributions. This fact does not prevent the hologram B from recording an image of C as it appears when illuminated by A, but, if C is displaced between two exposures, then the path length differences cannot be defined because every point at C is illuminated by rays arriving from different directions. Thus no well defined fringes will be recorded at B.

If A is non-scattering, or a point source, then every point at C will be illuminated by well defined rays from the point of illumination. When C is displaced, one single path length difference can, however, still not be defined over the total area of B, because the rays travelling from a point at C towards B will have to pass along different path lengths depending upon whether they reach the centre or a corner of B.

If, on the other hand, both A and B are point-like, or at least very small, then any type of displacement of C will, for each point on its surface, produce a well defined path length difference for light from the point A to the point B. Thus hologram interferometry can be used in a general way on randomly scattering objects only if illumination and observation are made from points or sufficiently small areas. When the virtual image is studied by a camera, as in Fig. 4.3, then the hologram plate is used only as a window and the aperture of the camera lens represents the observation area B which has to be sufficiently small. If, instead, the real image is formed by illuminating the hologram plate by a beam that is antiparallel to R (Section 2.3.6), then the illuminated area of the plate represents the observation area B. Consequently, the fringes on the object are defined only if the reconstructing beam is sufficiently thin.

Thus, in the general case, only one large scattering area is allowed in the practical use of hologram interferometry. So what if C is also non-scattering, e.g. a mirror? If A is a point source and if B is a point of observation, then only one single point will be seen on C and thus no fringes can be counted. Had A been a large scattering surface instead, then this surface would be seen reflected in C and the path length from A to B via C would be well defined so that, for example, a tilt of C would produce high quality fringes at its reconstructed image.

Finally, consider the situation in which B is a large area while A is again made a point source and C remains a mirror. In that case the point of illumination (A), as seen reflected in C, will grow into a large area. This situation could be realized by forming the real image of C by using a beam that is antiparallel to R. If this reconstruction beam is wide enough to cover the whole plate B, then this total area will scatter the light towards the image. As the different parts of the hologram plate remember the point A reflected in the mirror C from different angles, the real image will show not only a point but an area reflected in C. Exactly the same result could be produced by photographing the virtual image of C by using the camera as in Fig. 4.3, but with such a large camera lens that it would cover the whole plate.

Thus, under the most general conditions, it has been shown that *not more than one large scattering area* (e.g. C) *is allowed when hologram interferometry is used*. It has also been shown that *one large scattering area is needed* for the interference fringes to be seen.

There exists one situation when, at first glance, even the fulfilment of these conditions appears not to be enough to produce high quality fringes. That is when the object consists of a translucent, scattering material such as china, enamel, some plastics, bone, teeth and other organic tissues. Such materials do not have a well defined optical surface. Light rays are reflected at different depths and therefore these materials produce the same result as if they were made up from two or more layers of scattering surfaces. To make holographic measurements possible such objects have to be painted so that they become opaque.

4.2 Trigonometric calculations

Consider Fig. 4.4. Let us study how the path length from the light source (A) to the observation point (B) via the object (C) changes as the object is displaced from C to C'. The displacement (d) is infinitesimal (a few wavelengths) compared to the distances AC and CB (hundreds or thousands of millimetres). The path length of the illumination is shortened by the distance l_A, while the path length of the observation is shortened by l_B. The number of fringes caused by the shortened path length is n. Thus:

$$l_A = d \cos \beta, \qquad l_B = d \cos \gamma;$$

$$n = \frac{l_A + l_B}{\lambda} = \frac{d(\cos \alpha + \cos \gamma)}{\lambda}; \qquad (4.2)$$

hence

$$d = \frac{n\lambda}{\cos \alpha + \cos \gamma}. \qquad (4.3)$$

Equation (4.3) is fundamental to hologram interferometry and shows how the displacement (d) is calculated from the number of fringes (n) found between the displaced point and a fixed point on the object.

4.2.1 The fringe cone in the direction of displacement

Let us assume that the direction of the displacement is known. The object point (C of Fig. 4.4) could, for instance, be positioned on a membrane, of which it is known that the displacement is perpendicular to its surface. In that

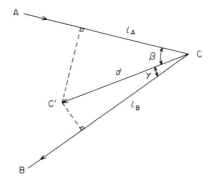

Fig. 4.4. Light arrives from A at the object point C and from there is scattered to the point of observation B. C is displaced by a distance d in the direction towards C'. The path length shortening will be $l_A + l_B$ if d is infinitesimal compared to the distances CA and CB.

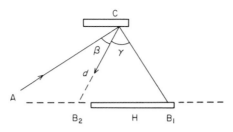

Fig. 4.5. The number of fringes seen between a fixed point and the displaced point (C) of an object will be at a maximum when it is studied in the direction of the displacement d. Thus the maximal number of fringes would be observed from B_2 if this point was within the area of the hologram plate. A is the point of illumination; H is the hologram plate studied from the observation points B.

case the angles β and γ are known and therefore the displacement (d) can be calculated using equation (4.3). However, if the direction of displacement is unknown, then it would be impossible to calculate the displacement from the one single reading of the fringes on the object. By changing the point of observation behind the plate the problem can, at least theoretically, be solved in a general way.

Let the hologram plate be of infinite size, so that the observer can always find the direction of displacement by moving his eye around behind the plate until he sees the object covered by the largest number of fringes between the displaced point and a fixed point (see Fig. 4.5). When this direction of observation (CB$_2$) that produces the maximal number of fringes is found, the line of sight towards the object point under study will also represent the line along which its displacement occurred ($\gamma = 0$).

The reason for the above statement is that in this case β is kept constant while γ is altered until the projection of d on CB is at a maximum. For each value of γ there will exist a certain number of fringes on the object and thus this fringe number will remain constant as the observing eye is moved along the surface of a cone having its top at the displaced object point and its axis parallel to the displacement. If the centre of this set of fringe cones is outside the hologram plate, the direction of displacement can still be found by calculating the motion of the point of observation, as described in the following section.

4.2.2 Moving the point of observation

As soon as the eye is moved away from the axis of the fringe cones the number of fringes will decrease. If the object is that of Figs 3.6 and 3.10, in which C represents the top of the bar, then fringes will be seen moving upwards and passing C as soon as the point of observation is removed sufficiently far from the direction of displacement. Let us assume that C is bright and that there are seven fringes on the bar when it is studied along the line of displacement. At a certain observation angle (γ) away from that line, C becomes dark because one dark fringe is moving upwards past the end of the bar. The situation will be the same for any direction of the angle γ and thus the top of the bar will look dark when it is studied from any point on a circle centred around the line of displacement. Further out, at a larger angle, C will again appear bright, but this time there will be only six fringes on the object.

Thus there will exist a set of concentric circles all centred along the direction of displacement. These circles, of course, represent the intersections of the plate by the cones of constant γ (Fig. 4.4). As an observer moves his eye behind the plate, the point which he is examining on the object will alternately appear bright and dark when his line of sight intersects the circles. By counting the number of cycles per length of eye motion it is possible to evaluate the displacement *even without being able to count the number of fringes on the object*. Perhaps the fixed point on the object is hidden or does not even exist at all.

The reason why an object point appears dark is, as stated earlier, that it has moved in such a way, between two exposures, that the image-carrying fringes on the plate have been blurred out. Thus the change in path length has, in the dark regions, been half a wavelength or an odd number of half-wavelengths. But this is exactly the same situation as was described for the Young's fringes caused by two points of illumination (Fig. 1.7).

The rings described above are therefore formed from the intersection of the hologram plate by the hyperboloids of the holo-diagram (Fig. 1.24(b)). These hyperboloids are rotationally symmetric around an axis in the direction of the displacement, and its focal points are the points under scrutiny before and after the displacement, respectively. From this reasoning it can be understood that what have previously been referred to as cones of constant fringe number are not really cones but rotationally symmetric hyperboloids viewed close to the axis. The cone approximation is, however, an extremely good one as the separation of the focal points of the hyperboloids is infinitesimal (a few wavelengths) compared to the distance to the plate (some hundred or so millimetres).

In the following a method will be described by which these intersections of the hologram plate by the hyperboloids can be viewed directly, without having to move the point of observation behind the plate.[9]

Let us form the real image on a cardboard screen by reconstructing the hologram using a laser beam that is antiparallel to the reference beam used during recording. No fringes are seen on the image because it is illuminated by the whole area of the hologram plate, of which different parts reconstruct the fringes at different locations on the object. Make a small hole in the cardboard at a certain point on the image of the object and, from behind, look through this hole towards the plate. Now only those parts of the plate will be seen as bright that diffract light to the object point where the hole is. Thus, through this hole it is possible to see the fringes on the plate that are identical to those Young's fringes formed if the object point existed in its positions before and after the displacement simultaneously.

4.2.3 The direction of maximal sensitivity

Let us now move on to solve a practical problem using the knowledge already gained. First, return to Fig. 4.4 and equation (4.3). Assume that the direction of displacement is unknown. In that case the illumination angle (β) and the observation angle (γ) cannot be calculated and it is therefore impossible to find the displacement from one single reading of the fringes on the object. The sum $\beta + \gamma$ can, however, be found from the holographic configuration. Let this sum be 2α (see Fig. 4.6). Thus

$$d = \frac{n\lambda}{\cos\beta + \cos(2\alpha - \beta)} \qquad (4.4)$$

or

$$n = (d/\lambda)[\cos\beta + \cos(2\alpha - \beta)], \qquad (4.5)$$

where n is the number of interference fringes caused by the displacement (d), while λ is the wavelength.

To find the value of β that produces the largest number of fringes (n), let d be constant and set $dn/d\beta = 0$, where $dn/d\beta$ is the change in n with respect to β. Then, differentiating equation (4.5), it is found that

$$\frac{dn}{d\beta} = \frac{d}{\lambda}[-\sin\beta + \sin(2\alpha - \beta)] = 0. \qquad (4.6)$$

Thus $\sin\beta = \sin(2\alpha - \beta)$ and therefore $\beta = \alpha$.

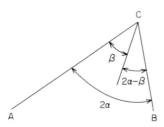

Fig. 4.6. It can be proved mathematically that the direction of displacement which produces most interference fringes bisects the angle ACB, where C is the displaced object point, AC the illumination direction and CB the direction of observation.

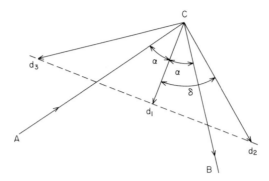

Fig. 4.7. The direction of increasing sensitivity to displacement bisects ACB. Thus the displacements d_1, d_2 and d_3, which all have the same projections on the bisector Cd_1, will result in identical number of fringes. C is the displaced object, AC the illumination direction, and CB the direction of observation.

This procedure shows that the direction of displacement which produces the largest change of the total path length is the one corresponding to $\beta = \alpha$. In other words, the direction of maximum sensitivity is along the bisector of the angle ACB. For this direction equation (4.3) is transformed as follows:

$$d = \frac{n\lambda}{2\cos\alpha}. \tag{4.7}$$

Figure 4.4 is transformed into the form shown in Fig. 4.7. Displacements in a direction that is perpendicular to the bisector produce no change of path length because the distance AC will be lengthened by the same amount as BC will be shortened. Thus, when there is a displacement (d) in a random direction only that component of d that is parallel to the bisector will have any influence on the path length. The three different displacements of Fig. 4.7 therefore all produce the same number of fringes. Transforming equation (4.7) so that it can be used for any direction of displacement (Fig. 4.7),

$$d_2 = \frac{n\lambda}{2\cos\alpha}\frac{1}{\cos\delta}. \tag{4.8}$$

If the movement is along the bisector (where $\delta = 0$), then the displacement needed to produce one fringe is

$$d = \frac{\lambda}{2\cos\alpha} = k\lambda/2. \tag{4.9}$$

The value of k is identical to the k-value of the holo-diagram and d is equal to the distance separating two adjacent ellipses of the diagram. This distance, which, of course, is measured perpendicular to the ellipses (and thus parallel to the hyperbolas) bisects the angle separating the beams of illumination and observation. The line $d_3 d_1 d_2$ of Fig. 4.7 is part of an ellipse which appears straight because the distances AC and CB are large compared to the displacements.

4.2.4 Calculating the true direction and amplitude

As seen in Fig. 4.7, and also from equation (4.8), one reading of a hologram is not enough to evaluate a totally unknown displacement. However, if it was possible to study several projections of the displacement on to bisectors in different directions, then it would be possible to deduce both its magnitude and its direction.

It is rather complicated to change the angle of illumination, but the angle of observation can be varied simply by studying the image through different parts of the holographic plate. If the observer moves his eye behind the plate so that the angle of observation is changed by $\Delta\alpha$, then the bisector will move through an angle $\Delta\alpha/2$. The result will be that different numbers of fringes are seen on the reconstructed image of the object when it is studied from different observation angles through the plate.

Let us start a practical evaluation by studying the holographic set-up as seen from above. Thus Fig. 4.8 is a vertical view of the holographic set-up,

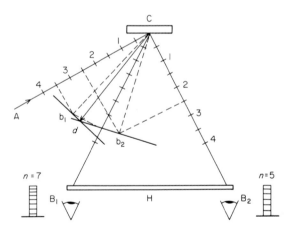

Fig. 4.8. Displacement evaluation by geometrical construction. C is the displaced object point; AC is the direction of illumination; CB_1 and CB_2 are two different directions of observation through the hologram plate H. Five fringes are observed from B_2, thus 2.5 units are marked along CB_2 and 2.5 units along CA. Normals are drawn through these points and their intersection marked at b_2. Seven fringes are observed from B_1, thus 3.5 units are marked along CA. Normals are drawn through these points and their intersection marked at b_1. Draw a normal to Cb_2 through b_2 and another normal to Cb_1 through b_1. The intersection of these two normals represents the true direction and magnitude of the required displacement (d).

where A represents the illuminating rays arriving from the laser. The object C represents the vertical bar of Fig. 3.6, as seen from above, while H is the hologram plate. When looking through the plate at B_1, the bar is seen as in the photograph shown in Fig. 3.10, but covered by seven fringes between the top and the fixed lower end. As the observer moves his eye from B_1 to B_2 the number of fringes decreases from seven to five. To find the direction and amplitude of the displacement at the top of the bar (C), that has caused these fringes, one proceeds as follows.

If the displacement that produced the seven fringes at B_1 had been solely directed along the bisector of the angle ACB_1, then the light path would have been shortened by 3.5 wavelengths along AC and by the same amount along CB_1. Let one wavelength be, for example, 1 cm and measure 3.5 cm from C along both CA and CB_1. Draw normals to AC and CB_1 respectively through these points. Where the two normals intersect, the point b_1 is found to which the arrow representing the displacement of C should be drawn if it had been along the bisector Cb_1. Draw a line perpendicular to Cb_1. Any displacement from C to this line would produce seven fringes when the object is studied from B_1. (Remember that Cb_1 is infinitesimal in relation to AC or CB.)

Now repeat the procedure using the information from B_2 about the five fringes. Measure 2.5 cm from C along CA and also along CB_2. Once again draw normals to AC and CB_2 respectively through the points. Where the normals intersect, the point b_2 is found to which the arrow representing the displacement of C should be drawn if it had been along the bisector Cb_2. Draw a line perpendicular to Cb_2. Any displacement from C to this line would produce five fringes at B_2.

The two lines that are perpendicular to the bisectors of ACB_1 and ACB_2 intersect at d. This intersection represents the only point that satisfies the conditions at both B_1 and B_2. Thus the unknown displacement has been found. Its direction is Cd and its magnitude is found by measuring the length of Cd in centimetres and multiplying this value by the wavelength.

4.2.5 Separating the out-of-plane from the in-plane motion

To evaluate the displacement two areas of information have been used. First, the area on the object was used to count the fringes between the displaced point and a fixed point. Second, the area on the hologram plate was used to vary the angle of observation to study how the fringe pattern changes. In this way it was possible to find, in a general way, the true direction and magnitude of displacement. Sometimes, however, it is advantageous to use a method that directly separates the displacement components that are in-plane and out-of-plane, respectively. In the following it will be seen that this too is possible by a calculated combination of the two areas of information just described.

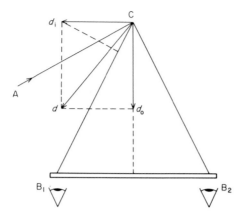

Fig. 4.9. The result of Fig. 4.8 is redrawn to get a clean figure for further analysis. The displacement (d) evaluated previously is divided into its components of in-plane (d_i) and out-of-plane (d_o) motions. The projections of d_i on CB_1 and on CB_2 are both 1 unit long, thus confirming the difference of two fringes between the observations from B_1 and B_2 respectively.

The result of Fig. 4.8 is redrawn in Fig. 4.9, where, however, the lines used to evaluate d are eliminated in order to get a clean figure for further analysis. The displacement (d) is divided into the two components that are in-plane (d_i) and out-of-plane (d_o). (Strictly speaking these two terms refer to the components that are perpendicular to and parallel to the line of sight respectively. The value of the angle B_1CB_2 is 2α and its bisector is normal to the object surface and also to the plate surface. The out-of-plane component d_o, which bisects B_1CB_2, produces exactly the same change in path length for ACB_1 and ACB_2. Thus d_o causes no difference in fringe number when observation is changed from B_1 to B_2.

The in-plane component d_i, on the other hand, produces changes in the path length that have different signs for the two observation angles. The projection of d_i on to CB_1 will be $d_i \sin \alpha$. Thus the path length ACB_1 will be shortened by this distance while ACB_2 will be lengthened by the same amount. The total difference in path length will be $2d_i \sin \alpha$. This value, when divided by the wavelength, is equal to the difference $n_1 - n_2$. Thus

$$d_i = \frac{(n_1 - n_2)\lambda}{2 \sin \alpha}, \tag{4.10}$$

where d_i is the in-plane displacement while n_1 and n_2 are the numbers of fringes seen on the reconstructed object when observed from B_1 and B_2 respectively. The number of fringes seen passing a certain point on the object when the eye is moved from B_1 to B_2 will be $n_1 - n_2$.

In Fig. 4.8 it has been assumed that $n_1 = 7$ while $n_2 = 5$ and therefore

$n_1 - n_2 = 2$. To test the correctness of the calculation, draw a normal through the head of d_i to CB_1 and measure the distance to C. The distance is 1 cm, indicating the total path length difference that caused the two fringes.

Thus we have found the following important relations:

(1) The fringes seen on the object surface between a fixed point and the displaced point represent the number of ellipsoids crossed by that point, the two focal points being the points of illumination and observation, respectively:

$$d = \frac{\lambda}{2 \cos \alpha} = k\lambda/2, \qquad (4.9)$$

where d is the separation of the ellipsoids, λ the wavelength, and 2α the angle between illumination and observation. Both the in-plane and the out-of-plane motions can contribute to the number of fringes.

(2) The fringes seen passing a certain point on the object as the point of observation is moved from one point to another represent the number of hyperboloids crossed by the studied point, the two focal points of the hyperboloids being the two points of observation:

$$d = \frac{\lambda}{2 \sin \alpha} = k^*\lambda/2, \qquad (4.11)$$

where d is the separation of the hyperboloids, and 2α is the angle between the two points of observation as measured from the object point under study.

It is interesting—and satisfying—to notice that the displacement normal to the line of sight (normal to the hyperboloids) found in this way is (apart from the constant of value 1.22) identical to the resolution of a telescope using the area of the hologram plate as aperture (see equation (1.11)). The transverse displacement needed to produce a fringe motion of one fringe also represents the smallest possible displacement that can be resolved in the form of a double contour in the reconstructed real image. Thus this value is, in a general way, fundamental to optics and not restricted to the studied case of hologram interferometry or to holography at all.

When the preceding calculations are applied to the special case of Fig. 4.9, the following simple relations are valid:

$$d_i = n_i k^* \lambda/2, \qquad (4.12)$$
$$d_o = n_o k \lambda/2, \qquad (4.13)$$

where n_i is the number of fringes passing the point under examination as the observation point is changed while n_o represents the number of fringes found on the object between the point under study and a fixed point.

However, if the assumed fixed point is not fixed, but rather the entire object has made an in-plane movement, then fringes will move over the whole object as the observation point is changed. From this movement it is possible to evaluate the rigid body motion using equation (4.12) and, by methods described in later chapters, to calculate how many of the fringes that are visible on the object are caused by the in-plane rigid body motion. The remaining fringes then result from the out-of-plane motion and can be evaluated using equation (4.13).

4.2.6 Evaluating the vertical component of motion

Until now we have studied only the two-dimensional case when the displacement is parallel to the horizontal surface of the laboratory table. If there had also been a component in the vertical direction the calculations would still have been correct, but we would not have been able to detect that component. The reason is that the path lengths ACB_1 and ACB_2 would not be influenced by a small vertical displacement of C if A, C, B_1 and B_2 are all along a horizontal surface. To find the vertical component it is necessary to study the fringe pattern on the object from observation points that are at different heights.

If the number of fringes is seen to be largest when an observer looks through the middle of the plate and decreases as the point of observation is lowered or elevated, in that case there is no vertical component of displacement. If, however, a difference is observed between the number of fringes seen through the top of the plate as compared to that seen through its lowest part, then the displacement has got a vertical component which can be found by using equation (4.12).

All the time, however, it must be kept in mind that, using double-exposure holography, it is not possible to find the sign of a displacement, only its direction and its magnitude. If, however, real-time holography has proved that the sign of Fig. 4.9 is true, i.e. that the path lengths have been shortened, then it is also possible to find the true sign of the vertical component. Should the number of fringes increase as the point of illumination is elevated, there is no doubt that the path length decreases and thus that the measured displacement had a vertical component directed upwards.

There is no problem in making the angle between the beams of illumination and observation (2α) small, e.g. 10°, and therefore the out-of-plane motion (along the line of sight) can be measured with high accuracy using equation (4.12):

$$d_o = \frac{\lambda}{2\cos\alpha} = 1.004 \times 0.5\lambda,$$

where d_o is the out-of-plane displacement needed to produce one fringe.

To find the in-plane motion (perpendicular to the line of sight), it is necessary to use equation (4.11), in which the angle between the two points of observation (2α) usually cannot be made larger than 45°, i.e.

$$d_i = \frac{\lambda}{2\sin\alpha} = 2.6 \times 0.5\lambda.$$

Often the angle has to be much smaller than this value. A plate size of 10 cm at a distance from the object of 1 m results in a value of d_i of only $20 \times 0.5\lambda$. From these results it can be concluded that if we did try to use equation (4.11) to calculate the out-of-plane motion, e.g. by moving the point of observation from B_1 to B_3 of Fig. 4.9, then we would throw away an appreciable part of the available resolution.

4.2.7 Young's fringes of observation

Previously we have thought of the hologram plate as only a "window with a memory" which is necessary for producing the interference effect, but otherwise has no influence on the calculations. Now we will move on to consider what happens during exposure and base our calculations on the changes of the speckle pattern recorded on the hologram plate. Let us re-examine the method introduced to explain how sandwich holography (Section 3.5) can be used to calculate tilt. The object is still the steel bar of Fig. 3.6. This time, however, it is not being bent but rather tilted about a horizontal axis at its lower end. Imagine the object being covered by a large number of randomly positioned mirrors. The directions of illumination and observation are along the line of displacement.

During the first exposure (Fig. 4.10(a)), one speckle ray is reflected horizontally from the top of the object (O) towards the hologram plate (H). After the first recording a force (P) is applied at the middle of the bar so that its upper end is tilted forward by an angle ϕ_1, after which a second exposure is made (Fig. 4.10(b)). The speckle ray is deflected downwards through an angle $2\phi_1$, as if it had been reflected by one of the small mirrors. Thus the intersection of the plate by the speckle ray will be lowered by a distance h:

$$h = L\sin 2\phi \sim L \times 2d/l, \tag{4.13}$$

where L is the distance between object and hologram plate and d is the displacement of the top of the object which has height l.

If, subsequently, the true real image of the object is reconstructed, it will be seen to be covered by Young's fringes caused by the double speckle points on the hologram plate used for the reconstruction. The separation of these fringes will be given by s, where

$$s = \lambda/2\sin 2\phi \sim \lambda l/2d. \tag{4.14}$$

The number of fringes (n) on the reconstructed object will be

$$n = l/s = 2d/\lambda. \tag{4.15}$$

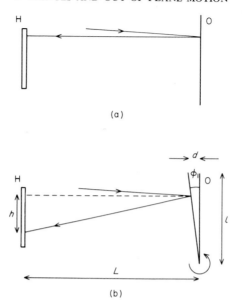

Fig. 4.10. (a) The mirror analogy to hologram interferometry. During the first exposure one speckle ray is reflected horizontally by the studied object (O) towards the hologram plate (H). (b) After the object (O) has been tilted through an angle ϕ_1, the speckle ray is deflected downwards through an angle $2\phi_1$ as if it had been reflected by a small mirror fixed to the object. During the second exposure the ray will hit the plate (H) at a point that is lowered by the distance h. The recorded twin speckles will diffract a laser beam passing through the plate towards the object so that Young's fringes are projected on to the object with a spacing that is identical to that of the holographic interference calculated in the conventional way (compare, for example, with Fig. 3.7).

This is exactly the same result as has frequently been derived previously when considering the path length difference:

$$d = n\lambda/2. \tag{4.16}$$

The true virtual image and the true real image are always identical if the same area of the hologram plate is used for the reconstruction and the reconstruction is carried out using an antiparallel (conjugate) beam. Thus exactly the same calculated fringe pattern will be found in the virtual image as in the projected real image. This fact can most easily be understood by using the concept of Young's fringes of observation,[1] which are caused by the interference of two spherical waves of observation, instead of the conventional Young's fringes caused by two spherical waves of illumination. This concept was introduced earlier (see Section 1.4.2).

4.2.8 The grating analogy

There is, however, one serious problem relating to the above method of deriving the fringe formation process from the concept of tilted mirrors. In all earlier calculations the angle of illumination had to be included, but when a mirror is tilted through the angle ϕ, its reflected beams are tilted through 2ϕ, independently of the angle of illumination. Therefore the method of replacing a speckle ray by the reflection from a small mirror is only relevant for certain special cases, e.g. the one shown in Fig. 4.10, in which the illumination and the observation are both made normal to the object surface and parallel to the displacement direction.

From this it can be seen that there are some serious limitations to the statement that a speckle ray behaves as if it was reflected by a mirror attached to the object. The mirror-to-speckle analogy is valid only if the mirror is parallel to the scattering object surface being studied. Thus the analogy is valid for a set-up like that in Fig. 4.11(a) but not for that in Fig. 4.11(b).

To study the motion of speckle rays in a more general way it is necessary to replace the mirror analogy by a grating analogy[2] (Fig. 4.12). A mirror parallel to the object OO would not reflect light from A to B, but a grating with the correct fringe spacing would. Let us see how the diffracted beam, which is analogous to the speckle ray, is influenced by a small tilt of the grating.

To simplify the reasoning let the distances to the illuminating point A and to the observation point B be infinite large compared to the grating spacing. Parallel light arrives from A at an angle α to the line normal to the grating (Fig. 4.12). It leaves, still as a parallel beam, at an angle β. The separation of two adjacent lines on the grating is g. The path length difference between the

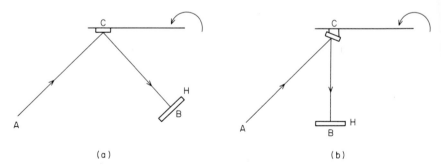

(a)   (b)

Fig. 4.11. The mirror analogy of Fig. 4.10 is strictly true only for scattering surfaces on which the angles of illumination and observation are identical. Thus the analogy can be applied to the set-up of Fig. 4.11(a) but not to that of Fig. 4.11(b). In the latter case the mirror analogy will produce a beam motion that is too large compared to the speckle motion it should mimic. C is the analogous mirror reflecting a laser beam from A to the centre (B) of the hologram plate (H).

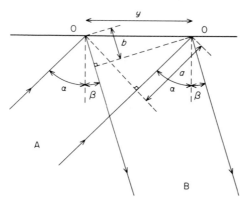

Fig. 4.12. When the mirror analogy does not apply, the grating analogy has to be used instead. The influence on the deflected light of a small tilt of a grating fixed to the object surface is examined. The separation of two adjacent grating lines is g. The light arrives from A at the angle α and leaves the grating at the angle β, travelling towards the point of observation (B). The path length difference for the two beams will be $a - b$ because the phase is constant along a line perpendicular to the direction of a light beam.

two diffracted rays will be $a - b$ because the phase is constant along a line perpendicular to the direction of a light beam. Thus we have the following equations:

$$a - b = \lambda, \qquad a = g \sin \alpha, \qquad b = g \sin \beta,$$

which result in

$$g (\sin \alpha - \sin \beta) = \lambda. \tag{4.17}$$

Therefore, considering a small change $\delta\alpha$ in α, which results in a small change $\delta\beta$ in β, it can be shown that

$$\delta\beta = \delta\alpha \frac{\cos \alpha}{\cos \beta}. \tag{4.18}$$

However, if $\delta\alpha$ has resulted from a small tilt of the grating through the angle $\delta\alpha$, then $\delta\beta$ represents the change in direction of the diffracted beam in relation to the grating. To get the absolute change of the angle β it is necessary to add the tilt of the grating. Thus

$$\delta\beta_{\text{abs}} = \delta\alpha \left(\frac{\cos \alpha + \cos \beta}{\cos \beta} \right). \tag{4.19}$$

Now we are ready to calculate the speckle motion (h) on the holographic plate at a distance L from the object:

$$h = \delta\beta_{\text{abs}} L. \tag{4.20}$$

The twin speckles recorded by the double exposure and separated by h at the hologram plate will produce Young's fringes of observation with the following separation (e) on the object:

$$e = \frac{\lambda}{2 \sin (\delta\beta_{abs}/2)} \sim \frac{\lambda L}{h}$$

$$= \frac{\lambda \cos \beta}{\delta\alpha (\cos \alpha + \cos \beta)}. \tag{4.21}$$

The number of fringes (n) seen projected on to the object image, which has length (l) and angle β to the line of sight, will be

$$n = \frac{l \cos \beta}{e} = \frac{l\delta\alpha (\cos \alpha + \cos \beta)}{\lambda}. \tag{4.22}$$

If the tilt of the object had been caused by a displacement (d) of one end, this displacement could be evaluated from $\delta\alpha$ of equation (4.22) in the following way:

$$d = l\delta\alpha = \frac{n\lambda}{\cos \alpha + \cos \beta}. \tag{4.23}$$

This equation is identical to equation (4.3), which was based solely on trigonometric calculation. Thus the use of the grating analogy combined with the concept of Young's fringes of observation produces a correct result independent of the angles of illumination and observation.

It is now time to enquire as to the necessity for seeking different ways to evaluate holographic fringes. One reason is that to carry out a certain calculation one particular method might be the most suitable, while for another calculation there might exist other methods that are easier to use. The last method presented produced a value for the speckle motion (h) on the hologram plate. Inserting this value in the formula for sandwich hologram tilt (equation (3.10)) produces an expression for the evaluation of sandwich holograms that is no longer limited to the spatial case in which the beams of illumination and observation have to be parallel to the direction of displacement. The resulting general formula is

$$\phi_1 = \frac{d}{Ln} \frac{1}{\cos \alpha + \cos \beta} \phi_2, \tag{4.24}$$

where ϕ_1 is the object tilt, d the separation of the emulsions, L the distance from the object to the hologram plate, n the refractive index of the emulsion, α and β the angles between the displacement and the directions of illumination and observation respectively, and ϕ_2 the sandwich tilt.

Later a formula similar to equation (4.19) will be used to compensate for the motion of the reconstructed image as the sandwich is tilted. The holographic image will be influenced by hologram tilt just as a diffracted ray is influenced by tilting a transmission grating. Equation (4.19) refers to a reflective grating; the corresponding equation for a refractive grating will be

$$\delta\beta_{abs} = \left(\frac{\cos\alpha}{\cos\beta} - 1\right)\delta\alpha, \tag{4.25}$$

where the symbols have the same meanings as in equation (4.19). The angle $\delta\beta_{abs}$ refers to the angular motion of a diffracted ray, or a speckle ray, or a reconstructed image.

If this term is included in the equation for sandwich hologram tilt (equation (4.24)), it will be found that the plate tilt necessary for fringe compensation becomes slightly larger because the image rotates with the plate. The difference, however, is so small that there is no reason to include this correlation factor when sandwich holography is applied to practical measurements.

One very important result gained by the combination of the grating analogy and the concept of Young's fringes of observation is that we now have a sound mathematical basis for understanding how sandwich holography works. We have calculated how much a speckle will move on the hologram plate (h of equation (4.20)). We also have calculated the Young's fringes projected on to the object by the set of twin speckles. It was demonstrated that spacing of these fringes (e of equation (4.21)) was identical to that of hologram interference fringes calculated in the conventional way (equation (4.23)). A manipulation of the separation of the twin speckles (h of equation (4.20)) will, of course, influence the result so that if h is made equal to zero there will be no fringes (e of equation (4.21) will be infinite).

4.2.9 The moiré pattern analogy as a consequence of the grating analogy

Let us go further with the study of the grating analogy to see if it might be useful in a more general way. During the first exposure a certain number of grating lines (n_1) along the object length (l) was needed to diffract light from A to B (Fig. 4.12). Subsequently the object was tilted through the angle $\delta\alpha$ and so the number of grating lines (n_2) must be different from n_1 if light is still to be diffracted towards B. In the following the difference in the number of grating lines ($n_1 - n_2 \sim \delta n$) will be calculated which is needed to diffract the light through the angle $\delta\beta_{abs}$ of equation (4.19).

Returning to equation (4.17) and setting the grating spacing (g) on the object equal to its length divided by the number of grating lines, it is found that

$$(l/n)(\sin\alpha - \sin\beta) = \lambda. \tag{4.26}$$

Thus, where $\delta\alpha$, $\delta\beta$ and δn are infinitesimal increments in α, β and n

respectively, it can be shown that

$$l(\delta\alpha\cos\alpha - \delta\beta\cos\beta) = \lambda\delta n. \tag{4.27}$$

Setting

$$\delta\alpha = -\delta\beta = \text{Object tilt} = d/l,$$

where d is the displacement of one end of the object which has length l, it is found that

$$d = \frac{\lambda\delta n}{\cos\alpha + \cos\beta}. \tag{4.28}$$

This equation is identical to equation (4.3), which was based on simple trigonometric calculations of path lengths.

Thus it has been proved that the number of hologram interference fringes seen on a tilted object is equal to the difference in the number of grating lines that are needed before and after tilting to diffract light from the light source to the observation point. The grating for each recording is simply found from the intersections of the object by the ellipsoids of the holo-diagram (Section 1.4.1).

In this way it has been demonstrated that the methods of Fig. 3.8 and Fig. 3.24 provide not only an approximate visualization but a mathematically correct analogy. In a later chapter (Section 5.5.2) it will be shown that a whole new field of interference fringe evaluation has been based on this method.

4.2.10 The vector approach

Any holographic interference pattern can be evaluated using simple trigonometry, e.g. equation (4.3). It is, however, time-consuming to make calculations point by point until the whole object has been studied from a number of observation points. Therefore several research workers have concentrated their efforts on simplification of the fringe evaluation.

Vector algebra and vector analysis are of course powerful tools for the processing of fringe data. The use of equations in the form of matrices facilitates automated handling of large quantities of vectorial information. In the vector method[3,4] the phase shift (δ) is calculated as the scalar product of the sensitivity vector ($\mathbf{k}$) and the displacement vector ($\mathbf{L}$), which is formally written as

$$\delta = \mathbf{k} \cdot \mathbf{L}. \tag{4.29}$$

The magnitude of $\mathbf{k}$ is $(4\pi/\lambda)\cos\theta$. Therefore the following calculation can be carried out:

$$\delta = 2\pi n = L(4\pi/\lambda)\cos\theta. \tag{4.30}$$

Thus

$$d = \frac{n\lambda}{2\cos\theta},$$ (4.31)

where n is the number of fringes while θ is half the angle separating the directions of illumination and observation, respectively.

This equation is identical to equation (4.7) where θ has been substituted for α and L has been substituted for d and the value $1/\cos\theta$ is identical to the k-value of the holo-diagram. Thus the vector $\mathbf{k}$ defines the direction and magnitude of sensitivity, while the k-value of the holo-diagram defines a desensitizing value. The direction of sensitivity is the same for the k-value (perpendicular to the ellipses) and the $\mathbf{k}$-vector (bisecting the directions of illumination and observation), but the numerical value of the one is proportional to the reciprocal of the value of the other.

It is, of course, unfortunate and confusing that the symbol k refers to two such different factors, both of which are used for hologram evaluation. The reason for these different approaches is probably that the k-value of the holo-diagram was introduced to solve the practical problem of finding the displacement that had caused a certain number of fringes. Thus a high k-value corresponds to a large displacement to fringe ratio. The $\mathbf{k}$-vector, on the other hand, was introduced to predict the fringe system that would be caused by a known displacement. Thus, a large $\mathbf{k}$-vector corresponds to a high sensitivity to fringe formation.

4.2.11 Regulated path length interferometry

It has been illustrated that by changing the angle of observation it is possible to find the true direction and amplitude of displacement. Thus, three observations through different parts of the hologram plate would reveal the three components of the displacement. To achieve a high accuracy the three observation points should be as far from each other as possible so that the difference in observation angle becomes large.

One disadvantage of this method is that the size of the hologram plate sets a limit on the size of distances separating the observation points. Even if the difference in observation angles can be made sufficiently large, e.g. by using several hologram plates, there can still be difficulties. One problem will be that the object looks different from different directions, so that the fringes become difficult to identify (see also Section 5.1.2).

Thus it would be advantageous to use three different illumination angles instead of three different observation angles. However, from a practical point of view, it is very difficult to make three different double exposures on three different hologram plates while changing the direction of illumination between each double exposure. An elegant solution to this problem is the "regulated path length interferometer" introduced by Zoltan Füzessy.[5] The method is as follows.

The light from a laser is divided into three beams, each of which illuminates the object from a different direction. The path length of each beam is so adjusted that it differs from that of the others by more than the coherence length of the laser light.

There are also three reference beams that illuminate the hologram plate at different angles. Each reference beam is so adjusted that it has the same path length as one of the illuminating beams. Thus three holographic images are recorded on the plate, each one representing the object illuminated from a certain direction.

The different images can be reconstructed one at a time by using a reconstruction beam angle corresponding to each specific reference beam angle. Each image is a recording of the object illuminated by just one illumination beam because, although all the illumination beams are all on simultaneously, only one of them is coherent with its corresponding reference beam.

The first exposure is made using all three illumination beams and reference beams simultaneously. After the deformation of the object a second exposure is made in the same way. The plates are processed and the doubly exposed holographic images are reconstructed using any one of the three reconstruction angles. Each image represents the object being illuminated by one single illuminating beam. The three sets of fringe patterns thus formed can be used to evaluate the three components of the displacement vector, using any of the evaluation methods described in this book.

4.3 The string analogy

Before taking a closer look at the holo-diagram, let us study a simple analogy based on stretched strings representing light rays. There is no need to solve any trigonometric calculations because that is done automatically when the analogous configuration is identical to that used during the holographic exposure.

For the practical evaluation of holographic interference fringes the present author has designed a mechanically analogous computer.[6] Stretched strings represent the path lengths of the light rays that travel from the light source to the observer via the object. The magnitude and direction of the displacement in three-dimensional space is displayed directly in the form of a mechanical movement. The method can be used in a general way; it evaluates the object point by point and is therefore influenced only by the number of fringes, not by the type of fringe pattern. For this reason calculation of the displacement is independent of the actual type of motion, e.g. rotation, translation, deformation or extension. It also works independently of whether the fringes are caused by an out-of-plane or an in-plane motion. Very little mathematical or optical knowledge is needed for use of the method.

In Fig. 4.19 (see later), A is the point source of laser light (e.g. a spatial filter) illuminating the object C. B is the point of observation (e.g. the eye of the observer). H is the hologram plate placed somewhere between C and B. A reference beam also illuminates H. The distance between the hologram plate and the point of observation does not have any influence on the principle of interpretation. It is, however, a practical simplification to refer only to points on the surface of the hologram plate. Therefore, in the following the observations to be made are restricted to a volume very close to and almost in contact with the hologram plate. The hologram is exposed twice, and if the object between the two exposures has been deformed, the hologram image of its surface will seem to be covered by fringes. Each dark fringe reveals that the path length of the light from A to B via the object has changed by an odd number of half-wavelengths between the two exposures. To find the amplitude and direction in space of the displacement that has caused the fringes the following method has been found to be easy to use and also to understand.

4.3.1 Line of displacement previously known

In Fig. 4.15, the object C is a vertical bar fixed to the horizontal surface of a table. The spatial filter is at A, the observation point at B, the top of the hologram plate at H; between the two exposures the top of the bar is bent in a known direction. The displacement of the top of the bar is so small that its direction can be assumed still to be parallel to the table surface.

In the hologram image the base of the bar is bright and so is the top. When the image is studied through the left-hand top corner of the hologram plate, four dark fringes are seen between its base and its top. The lowest dark fringe represents a difference in path length of 0.5λ, the next bright fringe represents a difference of $2 \times 0.5\lambda$, the next dark fringe a difference of $3 \times 0.5\lambda$, etc. The requisite displacement of the top of the bar thus represents a difference in path length of $8 \times 0.5\lambda$. To transform this value into the amplitude of the displacement, we must proceed as follows.

Take a string and make some 20 knots near one end. Make the distance between the knots 1.5×10^4 times the wavelength of the laser light used for the hologram exposure. A He–Ne laser will give a knot distance of $0.6328\,\mu m \times 10^4 \simeq 10\,mm$. Fix the string to the top left-hand corner of the hologram plate, and fix a small ring to the top of the object. Let the string pass through this ring and from there to the spatial filters, where the string is made taut and the other end fixed. The string now represents the path length of the light during the first exposure. The path length difference was 4λ

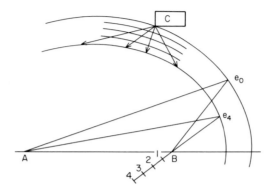

Fig. 4.13. A is the point of illumination; B is the point of observation; C represents a point on the object before deformation. C is situated on the ellipsoidal shell e_0 formed by the string Ae_0B, which represents the path length ACB. After deformation four dark fringes are seen on the object between C and a fixed point, indicating that the path length can now be represented by the ellipsoidal shell e_4. If only one observation is made, almost any vector from C to e_4 can represent the displacement.

between exposures. Loosen the ring from the top of the bar and shorten the string by four knots and again fix it to the left-hand corner of the hologram plate. The string now represents the path length of the light during the second exposure. The string is now too short to allow the ring to reach the top of the bar. If the ring is moved around while the string is kept stretched, it will travel on the surface of an ellipsoid (Fig. 4.13). At one point the known line of displacement intersects this ellipsoid. An arrow from the top of the bar to this point represents the direction and amplitude of displacement of the top of the bar. The direction was known and the amplitude is magnified 1.5×10^4 times.

Mathematically the point of intersection can be found by a system of two equations, one representing the line of displacement, the other representing the ellipsoid. In practice this point is found by moving the ring along the line of displacement until the string is stretched.

4.3.2 Plane of displacement previously known

In this case the direction of displacement is unknown; we only know that it is parallel to the plane of the table. Repeat the former procedure, but this time use the top right-hand corner of the hologram plate and assume that two dark fringes are counted on the bar. Loosen the ring again, shorten the string by two knots, and let the ring travel along the string forming a new ellipsoid. The first ellipsoid, corresponding to the left-hand corner, the second ellipsoid, and the plane of displacement intersect at one point. An arrow from the top

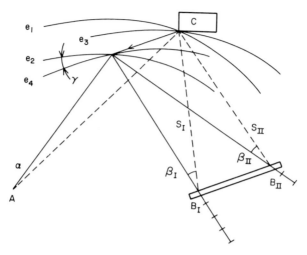

Fig. 4.14. Four fringes are seen on the object when the observation point B_I is used. Accordingly the string S_I is shortened by four knots, indicating that C has moved from the ellipse e_1 to e_2. Only two fringes are seen from B_{II}, indicating that C has moved from e_3 to e_4. An arrow from C to the intersection of e_2 and e_4 represents the displacement vector. The result is strictly correct only if the arrow is infinitesimal compared to the distances AC, CB_I, and CB_{II}. The sign of the displacement vector is not determined.

of the bar to this point represents the direction and amplitude of the required displacement (Fig. 4.14).

Mathematically, the point of intersection can be found by solving a system of three equations. Two equations represent the two ellipsoids; the third represents the plane of displacement. In practice the point of intersection is found in the following way. Fix two strings to the spatial filter, let them both pass through the ring fixed to the object, and let the strings go to the top left-hand and top right-hand corners of the hologram plate. Repeat the procedure of loosening the ring and shortening the strings as has been previously described. Thereafter, move the ring along the plane of displacement until both strings are simultaneously stretched. The ring is now at the point of intersection. This method should be compared to the trigonometric method of Fig. 4.8.

4.3.3 Nothing concerning displacement previously known

In this case it is necessary to study the fringes seen on the hologram image of the object using three corners of the hologram plate. Three strings are needed, and three ellipsoids are formed. An arrow from the top of the bar to the point of intersection of all three ellipsoids represents the amplitude and direction in

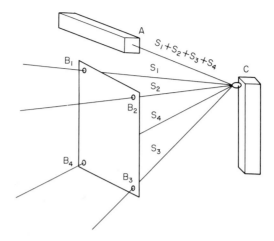

Fig. 4.15. The stretched strings (S) represent the straight light rays that jointly travel from the light source (A) to the studied point on the object (C) and from there split up and travel to the different points (B) of the hologram plate (H).

space of the required displacement. The amplitude is, as before, magnified 1.5×10^4 times.

Mathematically, the point of intersection can be found by solving a system of three equations, each of which represents one ellipsoid. To find the point of intersection in practice, the three strings are all put through the ring fixed to the bar. The strings are first made taut; then the ring is loosened and the strings are shortened according to the number of fringes seen through the corresponding corners of the hologram plate. The ring is then moved around in space until a point is found at which all three strings are stretched. This point represents the point of intersection. (Three of the four strings in Fig. 4.15 are used.)

4.3.4 No fixed point at the object

In this case one more quantity is unknown and thus one more equation is needed. All four corners of the hologram plate have to be used if the direction of displacement is completely unknown (Fig. 4.15). Assume that the base of the bar is hidden so that the total number of fringes on the bar cannot be counted. However, so much of the bar is visible that the direction of movement of the fringes can be studied. Assume also that an unknown number of fringes (n) exists on the bar when it is studied through the top left-hand corner of the hologram plate. Move the eye to the top right-hand corner of the plate while counting the number of fringes (q_1) that are added (moving downwards from the top of the bar) or subtracted (moving upwards and disappearing). If q_1 is positive, shorten the top right-hand corner string by q_1 knots. If q_1 is negative, lengthen the string by the same amount.

Move the eye downwards to the bottom right-hand corner of the hologram plate and again count the number of fringes (q_2) to be added or subtracted. Make the corresponding change of the string length. Proceed with the remaining corner and then move the eye upward again to the starting (top left-hand) corner and ensure that $q_1 + q_2 + q_3 + q_4$ adds up to zero.

We now have to find the point of intersection of four ellipsoids of unknown absolute radii but with a known relation between the radii and with known focal points. Mathematically, this corresponds to finding the solution to a system of four equations having four unknown quantities. In practice the solution is similar to the earlier examples. However, if we try to move around the ring with all four strings, we will not be able to find any point in space at which all four strings are stretched. The reason for this is that we have not changed the length of the strings by the right amount; only the relative change is correct!

Therefore clamp all the strings together near the spatial filter and let them slip together through a ring fixed to the filter. Fix the joint ends of the strings to a loaded spring. If we *now* move around the ring at the top of the bar, we will finally be able to find one point at which all the strings are taut. An arrow from the top of the bar to this point represents the amplitude and direction in space of the displacement wanted.

If the strings were also knotted at the end slipping through the ring fixed to the spatial filter, we would find that n knots have slipped through the ring. This number (n) represents the unknown number of fringes that exists on the bar as seen from the top left-hand corner of the hologram plate. This rather complicated method should only be used if absolutely necessary. When there is no fixed point on the object, it is often possible to introduce some sort of a bridge between the object and a fixed point. This bridge can, for example, consist of a metal strip or, as demonstrated in Fig. 4.16, the lever of a measuring instrument.

4.3.5 Practical uses

Strings, knots, slip-rings, and manipulation are probably not over-attractive to most scientists and engineers. However, the present author has found that, with appropriate equipment, the method is reasonably convenient to use. Sometimes the sensitivity of the methods described is too low, e.g. because the hologram is too small or is placed in an unfavourable position. Also the sign of the displacement cannot be found because the hologram plate does not remember which exposure was first. These two disadvantages are, however, caused by the holographic method and not by the method of evaluation.

Fig. 4.16. The holographic image of the vertical bar used in the experiment. At the top of the bar is seen a lever attached to an instrument used to measure the displacement. Observe that the identical number of fringes is seen on the bar and on the lever.

Some errors are still introduced by the string analogy, so let us study and compare the limitations of the method.

The accuracy of the string analogy described above is limited by the following four factors:

(1) The limited information content of the hologram. To get higher accuracy the position of the hologram plate has to be changed, or the hologram plate has to be replaced by a larger one, or two or more hologram plates have to be used simultaneously.

(2) Errors caused by the limited resolution in the reading of the fringes. The present author has found that in practice it is often difficult to resolve more than half a fringe, that is, half the distance between two dark fringes. This error, calculated as a percentage, is inversely proportional to the number of fringes.

(3) Errors caused by the elasticity of the strings. These errors can be kept to less than 0.5% of the string length.

(4) Errors caused by the amplification of the displacement vector. In actual hologram interferometry the displacement amplitude is of the order of light wavelengths, while the corresponding displacement in the string analogy has to be of the order of centimetres. The angles α, β_I, and β_{II} of Fig. 4.14 are so small that their influence can be neglected in hologram interferometry but not in the string analogy.

One way to make the string analogy strictly accurate would be to give the distances and the differences in distances equal magnification. The elasticity of the strings, however, would make that method impossible in practice.

A second method would be to give the strings a parallel movement by introducing some mechanical procedure.

The present author uses a third method which is based simply on accepting the errors and minimizing their effect. Because of the angles α, β_I, and β_{II}, the geometry of the string analogy during the second exposure no longer corresponds to the geometry of the hologram set-up. In one experiment, when the angle γ (Fig. 4.14) was 30°, it was found that to keep the amplitude error smaller than $\pm 10\%$ the displacement vector had to be less than 10% of the shortest of the distances AC, AB, or BC. In that case the direction error was less than $\pm 10\%$.

The following procedure drastically reduces the errors. First, lengthen the strings by the number of knots corresponding to the number of fringes and let the meeting point of the stretched strings represent one end of the displacement vector. Thereafter, shorten the strings by twice as much as they were previously lengthened and let the meeting point of the stretched strings represent the other end of the displacement vector.

The results of this procedure are that *on average* the angles become correct and that the errors are minimized. In the above experiment it was found that, when this method was used, the error of amplitude was less than $\pm 5\%$ and that the error of angle was less than $\pm 2°$ if the displacement was 10% of the smallest distance.

If, alternatively, a $\pm 10\%$ error in amplitude and a $\pm 10\%$ error in direction were accepted, a displacement of about 30% of the smallest distance could be tolerated.

One advantage of the method is that it is easy to understand. Very little prior mathematical or optical knowledge is needed. The taut strings represent the straight light rays that jointly travel from the spatial filter to the point on the object which is under study, and from there split up and travel to the different points of the hologram plate. The reason for using the corners of the plate is, of course, that we want to study the object from angles which are as different as possible.

The accuracy of the method can be assessed by hand. It is possible to feel whether the meeting point of the taut strings is well defined. It is also easy to see what directions of the strings would give higher accuracy and thus to find out how the position of the hologram plate should be changed for an improved experiment. The method also tells where to place a second

Fig. 4.17. Four stretched strings represent four light rays. The length of a string can be changed by turning the corresponding drum. The starting point of each string is determined by a slip-ring fixed to a hole representing the observation point of the hologram plate.

Fig. 4.18. Turning the knob by one step represents a displacement of half an interference fringe. The rotation of the corresponding drum changes the string length so that a measured displacement of 1 cm represents a real displacement of 1 μm.

hologram plate to increase the sensitivity. The angle between the intersecting ellipses (γ on Fig. 4.14) should be as large as possible. No pen or paper are needed. No calculations have to be made. The result is represented by a vector in space, which is where the engineer usually wants it, and not as coordinates on a strip of paper. The accuracy is high enough for most engineers' purposes, and all the interferometric information that exists on the hologram plate can be used. The method is self-contained and relatively quick.

4.3.6 Experiences

The present author has built a simple analogue computer using the method described. A simulated hologram plate-holder is equipped with four knobs, one at each corner (Fig. 4.17). Each knob is connected to a drum that can change the length of a corresponding string (Fig. 4.18). The starting point of each string is determined by slip-rings fixed to holes representing different observation points. Each knob is fixed to a cogwheel that gives 20 clicks in one revolution. When evaluating a hologram the first click represents a dark fringe, the second a bright fringe, etc.

Nature has generously arranged that the numerical value of the wavelength of a He–Ne laser is just a little more than 2π. The result of this is that a drum with a radius of 1 cm together with an appropriate string diameter will give a scale factor of 10^4. In the analogue machine 1 cm thus represents a real displacement of 1 μm.

The simulated spatial filter is equipped with four electrical contacts that give a signal when the force stretching the corresponding string has reached a certain value.

The device can be used either at the actual set-up to be utilized during the exposure of the hologram or in a special analogue set-up used only for reconstruction and evaluation. In the experience of the present author the latter method is much more useful.

It has been found that holograms with this device can be evaluated in less than 1 min for each object point under study, and the accuracy is limited mainly by the reading of the fringes and the sensitivity of the holographic set-up.

The accuracy can be checked by making a second evaluation with one reading deliberately made one fringe wrong. If this gives the result that the amplitude and/or direction of displacement changes by a large amount, the accuracy was obviously low and a change of the measuring conditions is recommended.

Very often the information content of a single hologram plate is not sufficient to give the required accuracy in the calculation of the three-dimensional displacement. In that case the described equivalence does, in a simple way, demonstrate the weakness of the holographic set-up; it also demonstrates how to improve the method, e.g. by using two or more simultaneously exposed hologram plates. The evaluation does not become any more complicated if the observation utilizes different hologram plates.

4.4 The holo-diagram

The holo-diagram was first introduced as a practical device for the making and the evaluation of holograms.[7] When used for this purpose it consists of the ellipsoids of constant path length from the focal point A to the focal point B as described in Section 1.4.1. Each ellipsoid represents a path length that differs from that of the adjacent ellipsoids by one wavelength. Thus the separation of the ellipsoids along the x-axis of Fig. 4.19 is 0.5λ while the separation is larger at all other places in the space around AB. The separation expressed in half-wavelengths is represented by the k-value, which is constant along arcs of circles (toroids), as described in Chapter 1. These k-values are printed alongside the y-axis in Fig. 4.19.

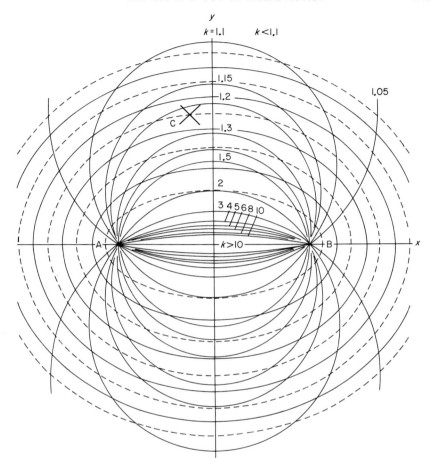

Fig. 4.19. The holo-diagram of Figs 1.21, 1.23 and 1.24(a) used for the evaluation of holograms. A is the point from which the divergent laser beam originates; B is the point of observation behind the hologram plate. Light from A to B via the object at C will not change its path length if C is displaced along an ellipse, while the difference in path lengths to adjacent ellipses is a constant number of wavelengths. The displacement perpendicular to the ellipses that is needed to cause one fringe is $k\lambda/2$, where k is constant along arcs of circles representing different spacings of the ellipses.

4.4.1 Point-by-point evaluation

To evaluate a displaced point on the object Fig. 4.19 can be used in the following way. Let A represent the point from which the laser light diverges to illuminate the object (strictly speaking it represents the pinhole of the spatial filter). B is the point of observation, the centre of the lens of the eye or the camera used for the observation. The hologram plate works like a window with a memory. Its position does not influence the fringe pattern as long as no optical component is changed between the two recordings of a double-exposure hologram, or between recording and observation of a real-time hologram. To simplify the evaluation, however, in the following let B be a point at the hologram plate, e.g. its centre.

Let the holo-diagram represent a vertical view of the horizontal table surface on which the holographic set-up is situated. Place the object into the diagram so that its size and its position in relation to A and B are equivalent to those of the real set-up. If there are n fringes between the point studied and a fixed point on the object, the displacement (d) is calculated:

$$d = nk\lambda/2, \tag{4.32}$$

where k is found by following the arc of the circle passing through the point studied and reading the k-value at the y-axis.

The value of d calculated in this way represents the smallest possible displacement that could produce n fringes, namely the displacement perpendicular to the ellipses. The true displacement could be in any random direction (Fig. 4.13); what has been calculated here is only its projection on to the normal to the ellipses (which is equal to the bisector of the directions of illumination and observation). Thus one reading from the holo-diagram represents just one measurement with the string analogy machine of Section 4.3. To retrieve the complete information concerning all three components of the displacement in space it is necessary to make three readings using different observation points.

4.4.2 Permissible approximations

Each fringe on the reconstructed image of the object is formed when the object point under study intersects one ellipsoid. Thus, the sensitivity depends both on the separation of the ellipsoids, the k-value, and the direction of the normal to the ellipsoids (which is parallel to the hyperboloids of Fig. 1.23).

The k-value is given by $k = 1$ along the x-axis outside AB, indicating that the sensitivity is at its highest possible value here; one fringe is formed for a displacement of 0.5λ along the x-axis. Between A and B the k-value is infinite, indicating that there is no sensitivity at all to displacements in any direction. This statement is true only *exactly* at the x-axis: as soon as we move away from it the k-value decreases. At large distances the asymptotic k-value decreases to a value $k = 1$.

The curvature of the ellipsoids will influence the fringe patterns. A large object that is close to A and B will have a sensitivity that varies to a great extent over its surface, while a smaller object far away from A and B will have a sensitivity which is approximately constant. In the latter case the fringe evaluation is simple; each fringe represents a displacement of 0.5λ in a direction parallel to the illumination and observation, and there is no sensitivity to displacements in any other direction. In that case the imaginary interference fringe surfaces are flat.

At closer distances, say less than ten times the separation between A and B, one must make an allowance in the calculation for the fact that the imaginary interference surfaces are no longer flat but can be approximated to spheres, their separation still being approximately constant at 0.5λ. At still closer distances, say within a sphere centred half-way between A and B and with radius AB, there exists a near field within which one must take into account that the k-value varies and that the imaginary interference surfaces consist of ellipsoids.

In this near field the holo-diagram is of its greatest value. Instead of having to study the direction of displacement and calculate its projections on to the direction of illumination and the direction of observation at every point on the object, the k-value is given directly in the holo-diagram and it is known that the maximal sensitivity is normal to the ellipses. In this way it is relatively easy to make at least a qualitative evaluation of the fringe pattern formed by a particular rigid body displacement of the object.

4.4.3 Translation at medium distance

If there exists no fixed point on the object, e.g. if it is translated in a direction perpendicular to its flat surface, then in that case no fringes are formed when it is at a large distance from A and B. However, if the object is somewhat closer to A and B, circular fringes are formed on its surface. Within the near field of AB still more complicated fringe patterns will form on its surface.

For the sake of simplicity, consider a flat surface placed at such a large distance from A and B that the distance AB can be neglected. Thus the ellipsoids of the holo-diagram degenerate into spheres that can be used to evaluate the situation shown in Fig. 4.20. Now consider the fringe forming in real time. In that case it will be seen that fringes form at the point of the object (C) that is closest to AB, because there the displacement (d) has the greatest influence on the path length ACB.

As d is increased, more and more circular fringes are found at C, after which they move outwards. The larger the displacement the more fringes are accumulated between C and a point at a certain radius (r) from C. If r is infinite, all the fringes will accumulate at the object. If r has a limited value, the fringes will pass the point at r with a lower rate than they are formed at C, and thus the number of fringes along r is a measure of d.

Let us study how the displacement (d) influences the path length (l) expressed in terms of the number (n) of wavelengths (λ) as described in Fig. 4.20:

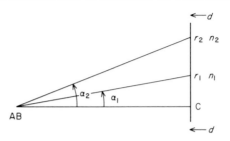

Fig. 4.20. A surface is illuminated from a point A and studied from another point B which is close to A. When the surface is displaced by a distance d towards A and B, circular fringes are formed. The fringe number n_1 is seen at radius r_1 and the fringe number n_2 at r_2. Equations are given in the test for the evaluation of d.

$$l = n\lambda = 2d \cos \alpha. \tag{4.33}$$

Thus

$$n = \frac{2d \cos \alpha}{\lambda}. \tag{4.34}$$

The different values of n found at two different radii r_1 and r_2 with corresponding α values will be as follows:

$$n_1 = \frac{2d \cos \alpha_1}{\lambda}, \tag{4.35}$$

$$n_2 = \frac{2d \cos \alpha_2}{\lambda}. \tag{4.36}$$

Thus

$$d = \frac{\lambda(n_1 - n_2)}{2(\cos \alpha_1 - \cos \alpha_2)}. \tag{4.37}$$

When the object is limited in size so that the centre of the circular fringes is not visible, it is necessary to count the fringes between r_1 and r_2 and insert this number as $n_1 - n_2$ in equation (4.37). If the centre is visible, then $\cos \alpha_1 = 1$ and the displacement is found by counting the number of circular fringes (n) and inserting that value in the following equation:

$$d = \frac{\lambda n}{2(1 - \cos \alpha_2)}. \tag{4.38}$$

From equation (4.33) it can be seen that there exists an angle α to every interference fringe. This simply means that the circular fringes seen on the object are intersections by a set of cones. From Fig. 4.21(a) it is seen that the fringes can be thought of as caused by the moiré pattern of two intersections

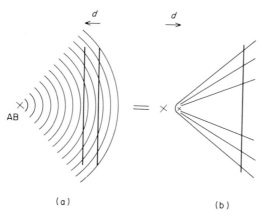

Fig. 4.21. Moving the object surface of Fig. 4.20 in the fixed set of spheroidal interference surfaces (with separation 0.5λ) is analogous to moving the set of spheres in relation to a fixed object. The moiré pattern thus formed will be a set of cones or, more strictly, of rotationally symmetric hyperboloids.

between the object surface and the spherical imaginary interference surfaces. Moving the object surface in the fixed set of spheres is analogous to moving the set of spheres that same amount in relation to the fixed object (Fig. 4.21(b)). The moiré pattern thus formed appears to be a set of cones or, more precisely, a set of rotationally symmetric hyperboloids, their two foci representing the observation points during the first and the second exposures as if they existed simultaneously. The analogous motion of the observation point should be double that of the object because the speckles recorded on the plate will move in depth along with the path length difference.

It must also be remembered that the evaluation methods described are true only if the points of illumination and observation are so close that the ellipsoids of the holo-diagram can be approximated to spheres. In all other cases it is necessary to study the more complex situation of the ellipsoids with their varying k-values and changing curvatures. In the next chapter it will be shown how this can be carried out automatically by studying the moiré pattern of the ellipsoids instead of the moiré pattern of the spheres, as in Fig. 4.21.

4.4.4 The near-field effects of translation in the x-direction

Now let us consider more closely the fringes formed on an object translated normal to its flat surface in the near field of AB.[8]

If the surface is perpendicular to the x-axis of Fig. 4.19, and thus the out-of-plane motion is parallel to the x-axis, then, because of the rotational symmetry, circular fringes will be formed. In the near field of AB these fringes will be influenced by changes in distances not only from A but also from B. Thus it is necessary to use more complicated equations than equation (4.34) and the approximations of Young's fringes of observation will no longer be acceptable.

To calculate the displacement in the x-direction that has caused a certain number of fringes it is necessary to invoke a new k-value (k_x) that is a measure, not of the perpendicular distance separating the adjacent ellipsoids, but of their separation in the x-direction. Thus we get the following equation:

$$c_x = \frac{2[y^2 + (L+x)^2]^{1/2}[y^2 + (L-x)^2]^{1/2}}{(L+x)[y^2 + (L-x)]^{1/2} - (L-x)[y^2 + (L+x)^2]^{1/2}}. \qquad (4.39)$$

The x-values of A and B are $-L$ and $+L$, respectively. In Fig. 4.22 the k_x-values have been calculated for the whole near-field space around A and B.

The loci of constant k_x-values also represent the loci of constant phase shifts caused by a translation in the x-direction. Thus the k_x contours also represent interference surfaces in space, and any object translated in the x-direction will be covered by fringes that are analogous to the intersection of the object surface by these k_x contours.

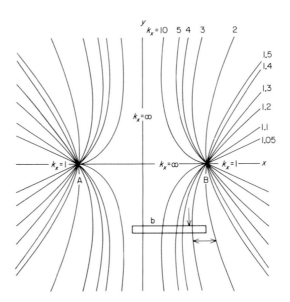

Fig. 4.22. If it is known beyond any doubt that the displacement is only in the x-direction, this diagram can be used. An object at b is translated in the x-direction. The displacement can be measured directly (in micrometres) by counting the resulting fringes on the object between the point where the k_x value is $2/\lambda$ ($k_x = 3.16$ for a He–Ne laser) and the y-axis (where the k_x value is infinite).

If it is known beyond any doubt that the only cause of the fringes is a displacement in the x-direction, then this displacement (d_x) can be calculated in the following way:

$$d_{x_1} = n_1 k_{x_1} \lambda/2, \tag{4.40}$$

$$d_{x_2} = n_2 k_{x_2} \lambda/2. \tag{4.41}$$

For a parallel translation $d_{x_1} = d_{x_2}$ and therefore it is possible to calculate d_x from the number of fringes ($n = n_1 - n_2$) on the object:

$$n = \frac{2d_x}{\lambda} \left(\frac{1}{k_{x_1}} - \frac{1}{k_{x_2}} \right), \tag{4.42}$$

where k_{x_1} and k_{x_2} are the k_x values of the extreme points of the object between which the n fringes have been counted.

The fringes are caused by differences in the sensitivity to fringe formation. If real-time fringes are examined, it can be seen how the fringes are formed at those parts of the diagram at which the sensitivity is at a maximum (low k-value) and, as the displacement increases, how the fringes move away from the point of maximum sensitivity towards the point of minimum sensitivity (high k-value). The fringes have to accumulate between the point at which they are formed and the point or line of zero sensitivity ($k = \infty$), which they are not allowed to pass. The fringes are most closely spaced where the k-value changes most rapidly, which is usually where the k-value is high.

If, for example, a long object is placed at b in Fig. 4.22 and displaced in the x-direction, the absolute amplitude of the displacement (measured in micrometres) is found by counting the number of fringes on the object in

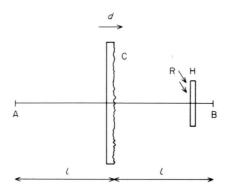

Fig. 4.23. A is the source of the diverging laser beam illuminating the ground glass (C) which undergoes a displacement (d) between the two exposures. H is the hologram plate which receives the reference beam R. The distance CH is shorter than CA. When the processed hologram is reconstructed no fringes are seen on the image of C when observed from a point B, so that CB is equal to the distance CA during recording.

between the y-axis and the point at which the k-value is about 3.16 (if the wavelength is 0.6328 μm). This type of interferometer can be read off by one count of the fringes, in spite of there being no fixed point on the object. Such an interferometer can be made in many ways; all that is essential is that different parts of the object surface have sufficient differences in sensitivity. The position described in Fig. 4.22 is, however, in many ways advantageous.

Figure 4.22 shows that k_x is infinite at the y-axis. This fact is also confirmed by the ordinary holo-diagram. All the ellipses are perpendicular to the y-axis and therefore a displacement in the x-direction will not influence the path length from A to B. To verify this statement the following experiment was carried out (see Fig. 4.23).

A flat, transparent, but diffusing surface (C) was positioned between the spatial filter A and the hologram plate (H). Between the two exposures of an ordinary doubly exposed hologram the object C was displaced towards H. When the hologram was reconstructed and studied from B no fringes were seen in spite of the fact that the displacement was not less than 4 mm (Fig. 4.24). However, as soon as the observer's eye was moved away from B, fringes appeared at the reconstructed object (Fig. 4.25). Thus the fringe-free situation existed only when AC was equal to CB so that C was exactly at the y-axis of the diagram shown in Figs 4.19 and 4.22.

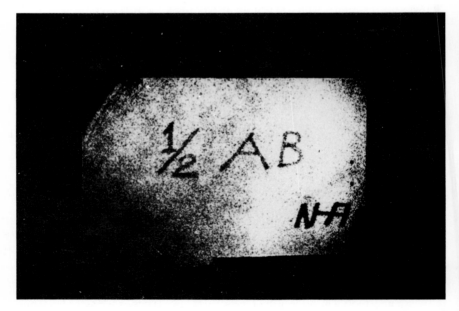

Fig. 4.24. The reconstruction of the ground glass (C) of Fig. 4.23 as seen from the observation point B so that CB is equal to CA. The displacement (d) was not less than 4 mm and still no interference fringes are seen in this photograph.

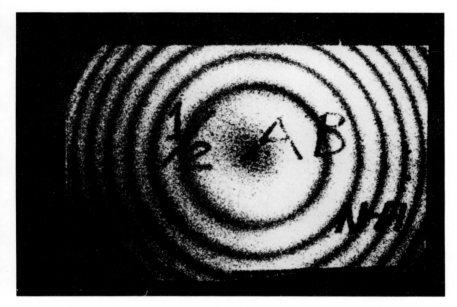

Fig. 4.25. Exactly the same reconstruction as in Fig. 4.24 but the point of observation (B) was moved forward so that CB became 30 cm shorter than CA. The displacement can be calculated by multiplying the number of fringes (seven) from the centre of the rings ($k_x = \infty$) to the right-hand corner with the k_x value at that point ($k_x \approx 2000$) and finally multiplying this value by half the wavelength of light. If the camera had been positioned exactly at the line of displacement, the centre of the rings would have been bright.

During the experiment the distance l was approximately 1 m (Fig. 4.23) and the camera was moved about 300 mm towards the plate between photographing the reconstructed images seen in Figs 4.24 and 4.25, respectively.

4.4.5 The near-field effect of translation in the y-direction

Let us now study the fringe pattern formed on a large object, e.g. a straight bar, placed parallel to the axis AB of Fig. 4.19 and displaced towards that axis. When the bar is far away from the x-axis a small motion towards AB will produce fringes that are formed at the y-axis and move outwards. The reason is that at large distances the k-value is almost constant but the centre of the bar moves almost perpendicularly with respect to the ellipsoids and thus it is highly sensitive to fringe formation.

When the bar is placed closer to AB the sensitivity at its centre decreases because the k-value becomes higher. At a certain distance the sensitivity will be almost constant along the bar. At still closer distances there will be two islands of maximal sensitivity opposite to A and B_1 respectively. Between

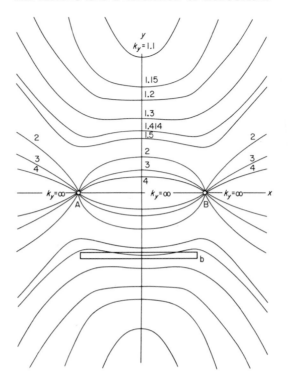

Fig. 4.26. If it is known beyond any doubt that the displacement is only in the y-direction, this diagram can be used instead of the original holo-diagram. Displacement (d_y) in the y-direction is calculated using the relationship $d_y = nk_y\lambda/2$. An object placed at b has a sensitivity to displacement in the y-direction that varies less than 2% even if that displacement is as large as AB.

these two points the sensitivity is lowered because of the high k-value. Outside these points the sensitivity is lowered because of the increasing angle between the normal to the ellipsoids and the displacement direction.

Figure 4.26 is similar to the ordinary holo-diagram, but the k-lines reveal the sensitivity to be in the y-direction instead of normal to the ellipses. Equation (4.43) has been used in the calculation of the diagram:

$$k_y = \frac{2}{y} \frac{[y^2 + (L+x)^2]^{1/2}[y^2 + (L-x)^2]^{1/2}}{[y^2 + (L+x)^2]^{1/2} + [y^2 + (L-x)^2]^{1/2}}. \tag{4.43}$$

The x-values of A and B are $-L$ and $+L$, respectively. The displacement in the y-direction (d_y) is calculated as follows:

$$d_y = k_y n\lambda/2. \tag{4.44}$$

This evaluation can, however, only be used if it is certain that the object has not been displaced in any other direction.

Let us study Fig. 4.26 where the x-values of A and B are $-L$ and $+L$, respectively. For a y-value of L, the k_y-value is $2^{1/2}$ at the y-axis, and varies little over the whole distance between x-values $-L$ and $+L$.

This line of relatively constant sensitivity to displacements in the y-direction can be very useful, as it can be used to simplify evaluation of the fringes on long objects. An object that has a length that is equal to the distance AB and a width that is a fifth of its length is positioned as demonstrated in Fig. 4.26. In that position it will have a sensitivity to translation in the y-direction that varies less than 2% over the whole surface. A tube with an inner diameter equal to $2L$ will, if its axis coincides with AB, have a sensitivity to radial movements that also varies by less than 2% over the length $2L$.

References

1. N. Abramson. Sandwich hologram interferometry. 3. Contouring. *Appl. Opt.* **5**, 200 (1976).
2. Lord Rayleigh. *Theory of Sound*. Macmillan and Co., London (1926), p. 122.
3. J. Sollid. Holographic interferometry applied to measurement of small static displacements of diffusely reflecting surfaces. *Appl. Opt.* **8**, 1587 (1969).
4. C. Vest. *Holographic Interferometry*. John Wiley, New York (1979), p. 70.
5. Z. Füzessy. Methods of holographic interferometry for industrial measurement. *Periodica Polytechnica, Budapest* **21**, 267 (1977).
6. N. Abramson. The holo-diagram. V. A device for practical interpreting of hologram interference fringes. *Appl. Opt.* **11**, 1143 (1972).
7. N. Abramson. The holo-diagram: a practical device for making and evaluating holograms. *Appl. Opt.* **8**, 1235 (1969).
8. N. Abramson. The holo-diagram. III. A practical device for predicting fringe patterns in hologram interferometry. *Appl. Opt.* **9**, 2311 (1970).
9. J. Gates. Holographic measurement of surface distortion in three dimensions. *Opt. Technol.* **1**, 247 (1969).

5

Evaluation of fringes caused by in-plane motion

Let us start this chapter with some philosophical views on the information content of holographic interference fringes before we get down to the more earth-bound problems of practical evaluation.

An ordinary map is a two-dimensional representation of three-dimensional space, the third dimension being the altitude. To visualize this third dimension contour lines are printed on to the flat map to display in discrete steps the altitude of the terrain at any point of the displayed area. The map could be explained as the terrain being intersected by a set of equidistant, flat, horizontal surfaces which results in the third dimension seen projected on to the plane of the flat map.

An ordinary doubly exposed hologram is a three-dimensional representation of four-dimensional space, the fourth dimension being, for example, displacements, deformations or vibrations. To visualize the fourth dimension interference fringes cover the three-dimensionally reconstructed object to display in discrete steps the displacement of the object at any point of its displayed surface. The interference pattern could be explained as the object being intersected by a set of more or less equidistant, flat, parallel surfaces. The result is that the fourth dimension is seen projected on to the volume of the reconstructed object.

The above analogy between a flat map and a three-dimensional hologram is quite a useful tool for understanding the basic evaluation methods. It does, however, break down when we study the fringe patterns in the near field of illumination and observation (Section 4.4.2) because there the imaginary interference surfaces intersecting the object are not equidistant, nor are they flat, nor are they even stationary. Their three-dimensional appearance depends on the angle of viewing—as if they were projected from a five-

156

dimensional space down to our three-dimensional space, in the same way as a three-dimensional cube looks different depending on the angle of observation when it is projected on to a two-dimensional screen.

No wonder that there are situations when the evaluation of holographic fringes in extreme cases becomes so difficult that methods other than mathematics have to be used. At the end of this chapter, as an example, an attempt is made to demonstrate how an analogous moiré pattern method can, in a simple way, be used to visualize intersections of the near-field interference surfaces.

5.1 Ordinary holography

If the above discussion about holographic interference fringes appears extremely complex, please, do not take it as a proof that the problem itself is difficult. *Holographic interferometry is not complicated in itself, it is the mathematical explanations that are complicated.* The true explanation of holographic interferometry lies simply in the changes of distances, as demonstrated, for example, by the string analogy of Section 4.3. The complexity is caused by the natural desire to use all the available information from the whole object at one glimpse of the pattern.

When an ordinary doubly exposed hologram is observed the reconstructed object is seen to be covered by interference fringes that carry discrete information about movements of the object. If the hologram plate is large enough, it can be seen that these fringes move and change pattern as the observer moves his eye. This method of studying the movement of the fringes, which is referred to as a dynamic observation, reveals information about the three-dimensional displacement of the object. Thus it can be said that there are two information-carrying areas in holographic interferometry: the reconstructed object surface is one area; the surface of the hologram plate is the other.

There are three principal methods for combining the information from these two areas:

(1) To use one point at the hologram plate for the observation of the fringe pattern covering the total area of the object: *static observation.*

(2) To study a single point at the object and observe how fringes pass this point when the eye is moved behind the plate until its total area has been utilized: *dynamic observation.*[1]

(3) To integrate the two areas of information by focusing the interference pattern on the total object by the use of, for example, a camera lens that is so large that it covers most of the hologram plate: *observing the fringe localization.*[2]

The first method is most sensitive to out-of-plane displacements whereas the second and third methods are most useful for the study of in-plane motions. Method 3 has some severe limitations as it can be used only when the motion of the fringes is such that it can be interpreted in terms of the parallax effect of a fringe system localized somewhere in space.

This chapter examines the three methods of hologram evaluation described above together with some special methods of recording that are especially designed for measuring in-plane motions.

5.1.1 Oblique illumination and observation

It has already been shown that the direction of sensitivity (the sensitivity vector) is perpendicular to the ellipsoids of the holo-diagram or, in other words, that it bisects the angle between the beams of illumination and observation. Thus, ordinary holography is highly sensitive to out-of-plane motion, because it is natural to place the object surface more or less normal to the plane of illumination and observation. It is impossible to have the surface of an opaque object parallel to the bisector because then the surface to be observed would be in shadow. Thus ordinary holography cannot produce fringes that are exclusively sensitive to in-plane motion (Fig. 4.1).

It is, however, possible to bring the direction of increasing sensitivity very close to the surface plane of a flat object by carrying out illumination and observation from oblique angles. In the preceding chapter the sensitivity was derived in the following way (equation (4.3) and Fig. 4.9):

$$d = \frac{\lambda}{\cos \alpha + \cos \beta},$$

where d is the displacement needed to produce a path length difference of λ, while α and β are the angles between displacement direction and the directions of illumination and observation respectively. For oblique illumination and observation the following results are obtained (Fig. 5.1):

$$d_{\mathrm{o}} = \frac{\lambda}{\cos \alpha - \cos \beta}, \tag{5.1}$$

$$d_{\mathrm{i}} = \frac{\lambda}{\sin \alpha + \sin \beta}. \tag{5.2}$$

If the object is flat, it is possible to make the illumination almost parallel to its surface. On the other hand, it is usual to keep the angle of observation as near as possible to being at right angles to the surface in order to get an undistorted view of the object. Inserting $\alpha \sim 90°$ and $\beta = 0$ produces $d_{\mathrm{i}} = d_{\mathrm{o}} = \lambda$, where d_{i} and d_{o} represent the in-plane and the out-of-plane displacements, respectively. For practical reasons the angle of illumination

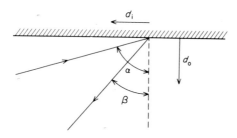

Fig. 5.1. To make a hologram with high sensitivity to in-plane motion (d_i) and low sensitivity to out-of-plane motion (d_o) the illumination angle α and the observation angle β should be as large as possible.

has to be slightly less than 90° with respect to the normal to the surface, and therefore this method has the large disadvantage that its sensitivity to out-of-plane movements is slightly higher than that to the in-plane motion which we want to measure.

To get around this problem it is necessary to give up the stipulation that the direction of observation be normal to the surface. In that case both α and β can be close to 90° and on the same side of the normal to the surface, so that d_i becomes close to 0.5λ while d_o becomes almost infinite. In practice α and β cannot be smaller than somewhere around 5°, which results in the values $d_i = 0.502\lambda$ and $d_o = 5.75\lambda$. Thus the proposed configuration produces a predominant sensitivity to in-plane motion. The most severe disadvantage is that the method can only be used on surfaces that are flat, otherwise some parts will be in shadow.

5.1.2 Image restoration

The fact that the surface of the object appears to be very foreshortened and distorted because of the oblique angle of observation can be overcome in the following way.[3]

During recording both the beam illuminating the object and the reference beam are collimated (Fig. 5.2). After a double exposure the processed plate is repositioned in the plate-holder and the true real image is formed using an antiparallel reconstruction beam (Section 2.3.6). This image will be focused in space exactly on the point at which the object was positioned during exposure, and thus the interference fringes will be projected on to the surface of the object—which remains in place during reconstruction. It is now possible to study the object with its projected fringes from any angle, and therefore the object can be photographed from a direction that is normal to its surface so that the object shape is not distorted by an oblique observation.

The in-plane displacement of any point on the surface is evaluated by

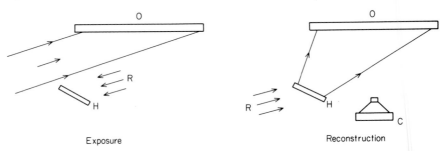

Exposure   Reconstruction

Fig. 5.2. Compensation for the distorted view caused by oblique observation. During recording both the illumination beam on the object (O) and the reference beam (R) on the hologram plate (H) are collimated. The reconstruction of the doubly exposed hologram is made using a beam that is antiparallel to the reference beam. The real image with its recorded fringes is thus projected on to the object surface where they can be photographed by the camera (C) in a direction that is normal to the surface.

counting the number of fringes between the point being studied and some fixed point and multiplying this value by d_i of equation (5.2). The main disadvantage of the method described here is that it can only be used on more or less flat surfaces. When the object surface is noticeably curved the angles of illumination and observation will vary so much that the technique becomes useless.

5.2 Difference between two images

In ordinary stereoscopy or photogrammetry two images taken at different angles and each representing the in-plane information are combined to produce depth, the out-of-plane information. In a similar way two interferograms taken at different angles and each representing depth, the out-of-plane information, can be combined to produce the in-plane information.

5.2.1 Moving the point of observation: the dynamic method

It has already been shown in Section 4.2.5 and Fig. 4.9 that in-plane motion can be evaluated from a hologram by counting the number of fringes passing a certain object point as the eye of the observer is moved behind the plate. Let us repeat equation (4.11):

$$d_i = \frac{n\lambda}{2 \sin \alpha},$$

where d_i is the in-plane displacement, n the number of fringes passing a certain object point, λ the wavelength, and 2α the change in the angle of observation.

The evaluation can be carried out in such a way that the eye is first moved vertically to find the vertical displacement component. Thereafter it is moved in a horizontal direction to find the horizontal component.

Another method of evaluation is to look for the angular direction of the eye motion that produces no fringe movement at the point under scrutiny. The in-plane displacement must then be in a direction that is perpendicular to this direction of zero fringe shift. The fringes at the object will be most sensitive to changes of observation along this perpendicular direction, and by counting these fringes the in-plane displacement can be found.

One great advantage with this method of measuring in-plane motion is that only the fringe shift is involved; information about the fringe order is not needed. Thus there is no need to have any fixed point at the object from which to count the fringes. One great disadvantage with the method is that its sensitivity is low, as was demonstrated in Section 4.2.6. The sensitivity is equal to the smallest displacement that can be measured using diffraction-limited resolution by observation through the area of the hologram plate and without the use of interferometry (Section 4.2.5). One more disadvantage of the method is that it does not, like ordinary holography, produce a map of the total displacement distribution. Instead the object has to be observed point by point, which makes evaluation rather time consuming. Attempts have therefore been made to automate the evaluation process, as described below.

What has been discussed so far is what the virtual image looks like when it is studied by eye through different parts of the hologram plate. Exactly the same result can be obtained by reconstructing the true real image using a reconstruction beam that is antiparallel to the reference beam used during recording (Section 2.3.6). To facilitate the method the latter should be collimated. Thus, by reconstructing with a thin laser beam scanning over the hologram plate, a real image with its surface covered by interference fringes can be projected on to a screen. These fringes move around as the reconstruction beam illuminates different parts of the plate. A photo-detector positioned on the screen at a particular point on the fixed image observes the variation in light intensity as the fringes pass by; its signal is used to calculate the displacement of the studied object point. The beat effect caused by the moving fringes can be evaluated using equation (4.11) because the evaluation is identical whether the information from a certain area of a hologram plate is revealed using the virtual or the real image.

Using this method it has been claimed that displacements of less than 0.1λ have been measured using electronics that are able to resolve a phase shift representing a hundredth of a fringe. Such a high resolution is difficult to reach in practical work. The experience of the present author is that variations in the hologram emulsion, and especially variations in the air caused by temperature gradients, usually produce errors in the fringe position of one-quarter of a fringe.

5.2.2 The combination of two images

To reach a high resolution the recording configuration should be such that it produces a large angle of observation even from large distances. The most obvious method is to use two, or more, hologram plates instead of just one. Returning to Fig. 4.8, it can be seen that a hologram plate at B_1 records seven fringes while a plate at B_2 records five fringes. The difference, two fringes, is caused by the in-plane component, which adds one fringe to the image at B_1 and subtracts one from that of B_2. The six fringes are caused by the out-of-plane motion which influences the path lengths to B_1 and B_2 in the same way. Thus it is not necessary to count the fringes that pass C as the eye is moved from B_1 to B_2. This would be impossible if the two points are on separate plates. It is sufficient to study the difference in fringe numbers between B_1 and B_2.

Production of a fringe pattern that is caused solely by the in-plane motion can be effected simply by putting the fringe pattern of B_1 on top of that of B_2. The resulting moiré pattern represents the difference between the two patterns. Therefore the identical pattern component caused by out-of-plane motion is eliminated and only the in-plane component remains. The evaluation of these moiré fringes is carried out using equation (4.11).

There are, however, two limitations to this method. One is that the two views must be identical, which means that the distortion of the object image must be restored, e.g. by the method of Section 5.1.2 in which the real image is back-projected on to the object surface, where it is then photographed.

Another limitation is that there must exist an out-of-plane motion, e.g. tilt, that produces more fringes than the in-plane motion to be measured. The reason for this is that the fringes resulting from the in-plane component should be added to the out-of-plane fringes in one of the holograms; in the other hologram they should be subtracted. As the sign of displacement is lost when the fringes are photographed, the addition and subtraction has to be made before the photo is recorded.

5.2.3 Two illuminating beams

There exists one simple way to get around most of the problems so far described, that is by illuminating the object obliquely from two sides while recording by the use of a single hologram,[4] as seen in Fig. 5.3. This set-up is, in principle, identical to that of Fig. 4.9, but the points of illumination and those of observation have been interchanged. Equation (4.11) is still valid, but now the angle 2α refers to the angle separating the two directions of illumination instead of the two directions of observation. One great advantage, however, is that the illumination angle does not influence the object image and therefore there is no need for image restoration. Another advantage is

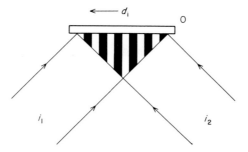

Fig. 5.3. Illumination by two beams (i_1 and i_2) results in fringes that are projected on to the object surface (O). If the object is doubly exposed using ordinary holography only those points on the surface will be recorded twice that were in a bright fringe during both exposures. Thus the method is sensitive to in-plane motions.

that the two sources illuminate the object simultaneously during the two exposures of the single hologram. One simple way to understand how this method works is as follows.

When the object is illuminated by the two coherent laser beams a set of Young's fringes are projected on its surface. The fringe separation (d) is (equation (1.3))

$$d = \frac{\lambda}{2\sin\alpha},$$

where 2α is the angle separating the two beams. The dark Young's fringes could be referred to as "fringe of blindness" as they carry no information. The object surface will be covered by these projected fringes which, however, are so closely spaced that they cannot be resolved by the hologram.

The points on the surface that happen to be in a dark fringe during the two exposures will, of course, have no influence on the recordings and thus no information about any displacements between exposures.

Those points that are displaced in such a way, between the two exposures, that they are moved from the darkness of one projected fringe into the brightness of its neighbour, or vice versa, will only be recorded once. Thus these points are prevented from delivering any information regarding displacement, so these areas will also be free from interference fringes.

All those points, however, that are in a bright projected fringe during the two exposures will be doubly exposed and will contribute to the formation of interference fringes. Therefore, ordinary holographic interference fringes will be seen on these areas, caused by the out-of-plane motion (or strictly speaking by intersected ellipsoids of the holo-diagram).

In this way the reconstructed object image is seen with its surface divided into areas that are alternatively covered by out-of-plane fringes or fringe-free.

Thus the fringe-free bands represent loci of constant in-plane displacement, evaluated using equation (4.11). In that case, however, the angle 2α should represent the angle separating the two directions of illumination instead of the two directions of observation. The direction of maximum sensitivity will be in the plane of the two illumination beams and normal to their bisector.

One disadvantage of the method described is that the surface under examination must be covered by fringes, e.g. by tilting the object or any optical component between the two exposures. This disadvantage is one which is held in common with the earlier method which uses one illumination beam in combination with two holograms. The greatest advantage of both methods is that they are insensitive to out-of-plane motions. One final disadvantage with the method described in this section is that it can sometimes be difficult to discriminate by eye between fringe-free and fringe-covered areas. In the next section a simple solution is presented to this problem.

5.2.4 Optical filtering

For quite a number of optical measuring methods the result is an image that at some areas is covered by fringes, while other areas are fringe-free. To discriminate between these areas one can use optical filtering, which transforms the areas with fringes and the areas without fringes into bright and dark areas, respectively. Such a transformation is especially useful when the fringes are so closely spaced that they are difficult to resolve.

First, it should be pointed out that the reconstruction of an ordinary transmission hologram is one example of optical filtering. Only those parts of the plate (the window with a memory) are seen as bright that are covered by a grating of such a frequency that light is diffracted from the laser to the observer. The dark interference fringes caused by displacement are examples of areas where the grating lines are missing because of a moiré effect between the two holographic recordings.

The term "primary fringes" has already been introduced for the image-carrying fringes on the hologram plate, and "secondary fringes" for the fringes of holographic interferometry. Thus higher-order fringes represent the difference between two low-order fringe patterns, and optical filtering is a method for enhancing the higher-order fringes while suppressing the lower-order ones. When a focused image hologram is studied it is evident that the primary fringes work like a carrier frequency that builds up the image. The reconstruction is studied by holding the eye in the first-order diffraction beam.

Figure 3.21 shows a good example of a pattern that would gain in contrast if it were optically filtered. The object is a bar covered by vertical lines that was vibrated from left to right to visualize the time-averaging effect. If a

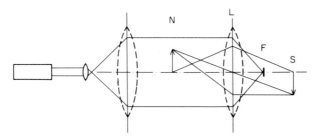

Fig. 5.4. Optical filtering. The negative (N) to be filtered is illuminated by parallel light and its image is focused by the lens (L) on to a screen (S). The parallel light passing directly through the focus (F) is stopped by a small opaque spot at the focus (F) of the lens. Thus only diffracted light passes this spot to produce a bright image at S.

negative of this figure is studied from such a direction that laser light is diffracted to the eye by the grating lines, then a secondary fringe pattern is seen representing time-averaged vibration fringes. This optical filtering phase is identical to the reconstruction phase of a focused image time-averaged hologram of a vibrating object, the reconstruction beam coming from one side.

To make the optical filtering process more effective the method illustrated in Fig. 5.4 is sometimes used. The image to be filtered is photographed and the negative (N) inserted in an imaging system. The collimated light from a laser passes through the negative, which is imaged on to a screen by the lens (L). If the primary grating on the negative is of sufficiently high frequency the first-order diffracted light will fall outside the illumination beam (the zeroth-order beam). In this way the required filtering is made automatically (holography). If the grating frequency is low the zeroth-order beam can be cut out at the focal plane of the lens by inserting a glass plate with an opaque spot exactly at the focal point. The light that is diffracted away from the zeroth-order beam will pass at the side of this stop. Thus an image is formed that consists of bright areas where diffraction takes place.

The filtering process can be made still more selective by exchanging the opaque spot for a pinhole in an otherwise opaque screen. Placing this hole at the first-order diffraction point of a grating, not only is the zeroth-order light filtered out but also diffracted light from other gratings with different frequencies or with different directions of their grating lines. Furthermore, those gratings that have the same direction but a frequency that is half, one-third, one-quarter, etc., of the required grating will also direct light through the hole in the form of second-, third- and fourth-order diffractions. Thus the method can select those areas on the image that have a primary fringe pattern that is of a particular direction and also has a particular frequency or multiple thereof.

If a number of frequencies is required rather than a single frequency, then the optical filter could have other more complicated forms than just the pinhole. To make the method more practically useful one or more lenses are added, but the basic principle of Fig. 5.4 remains.

5.3 Fringe localization

Let us return to the method of utilizing one large hologram plate close to the object and studying how the fringes move as the eye of the observer is moved behind the plate (Section 5.2.1). Sometimes this motion can be interpreted as a parallax effect caused by the fact that the fringes are localized somewhere in front of, or behind, the object surface.

If the fringes move to the right, relative to the object surface, as the observer's eye is moved to the left, then the fringes appear to exist somewhere in front of the object. If, instead, the fringes move in the same direction as the eye of the observer, then they appear to exist behind the object surface. If the fringes move in a vertical direction when the observer moves in a horizontal direction, then the fringe motion cannot be interpreted as resulting from any parallax effect at all. The same problem arises if the fringe pattern changes, e.g. if it is made up from a set of circular fringes the diameter of which changes as the viewing angle is varied.

There are at least three different ways to utilize the information provided by holographic interferometry. The first is to study the fringe pattern on the object through one point of the plate. The second is to study the fringe changes at one point of the object through different parts of the plate. The third is to mix the two sets of information from the area of the plate and the area of the object. The last-mentioned method means that by using a camera of large aperture placed close to the plate its information is integrated over a large area.

If the fringes, when seen from different parts of the camera lens, behave in such a way that their angular position can be explained in terms of the spatial existence of the fringes, then it is possible to focus the camera until the fringe contrast is at a maximum. By reading the focusing distance of the camera it is possible to establish the fringe localization—if such a localization exists. The present author has doubts about the usefulness of an evaluation method that is limited to those situations in which the motion of the fringes happens to be such that they appear to have a fixed spatial existence. The alternative procedure of examining the hologram point by point can lead to the utilization of all its existing information.

5.3.1 General aspects

Fringe localization has given rise to considerable interest within the research community as a means for the evaluation of displacements. In this and the following sections the function of fringe localization will be examined, but no attempt will be made to go into all the complicated mathematics that has been devised to explain the apparent localization of something that is not localized.

The fringes appear localized at infinity if the object surface is flat, is illuminated by parallel light and is translated without rotation. The reason for this occurrence is that there exists no specific feature that characterizes any one part of the surface with respect to the others. All points on the surface are equal except for one thing, the angle of observation. Therefore the fringes are only influenced by the point of observation, and the fringes move when this point is changed. In whatever way the point of observation is altered, each fringe will exist at a certain angle of observation, just as the image of the Moon does. Consequently these fringes appear to be localized at infinity.

On the other hand, if there exists an extreme point of motion somewhere on the surface of the object, then, of course, the fringes appear more or less fixed at that point. Thus, if a flat surface rotates around an axis normal to the plane of its surface, this stationary axis can never be passed by an interference fringe. When the angle of observation is varied the fringes can only rotate around this point. The resulting effect is that the fringes in the neighbourhood of the axis appear to be more or less localized at the surface. Usually the fringe pattern gives the impression of intersecting the plane of the surface along a line perpendicular to the fringes and passing through the object's axis of rotation.

When the axis of rotation lies in the plane of the surface, no fringes are allowed to pass this axis because it is not displaced. Therefore the fringes line up parallel to the axis and a change in the angle of observation commonly influences the fringe spacing.

If there is a complex motion, e.g. a combination of rotation, translation and deformation, the plane of rotation will be transformed into a curved surface of location with a topography that can become extremely complicated.

5.3.2 Homologous rays

Long before holography was invented there was an interest in the localization of interference fringes. The Newton's rings (Section 1.3.2) caused by two flat parallel sheets of glass separated by a thin film of air and illuminated by the cloudy sky will appear to be localized at infinity as there is no extreme point at the glass surfaces. The concept of homologous rays has been used to explain, for example, the geometry of the localization of the fringes in a

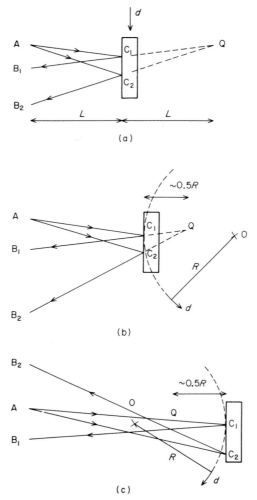

Fig. 5.5. (a) Localization of fringes caused by in-plane motion (d_i) of the object. During the first exposure light from the source A is reflected towards B_1 as if a small mirror were fixed to the object at C_1. Prior to the second exposure the same mirror has moved to C_2. The localization of the resulting fringes is at Q, where the two homologous rays B_1C_1 and B_2C_2 intersect. Thus the localization will be at the same distance behind the object as the illumination source is in front of it.

(b) The localization of fringes caused by rotation around the axis (O) behind the object surface. Prior to the second exposure the small mirror at C_1 is moved to C_2 because of a rotation around (O). The localization of the resulting fringes will be at Q, where the two homologous rays B_1C_1 and B_2C_2 intersect. Thus the apparent localization will be behind the object and approximately half-way to the axis of rotation.

(c) Rotation around an axis (O) in front of the object causes the resulting fringes to appear in front of the object, approximately half-way to the axis of rotation.

Michelson interferometer with a broad diffuse illumination source, as described in Section 1.2.4 and Fig. 1.18.

Homologous rays are rays from the same point on the scattering surface and the theory is based on the fact that fringes are formed only when a scattering point interferes with itself in two different positions.[5] Different scattering points do not interfere with each other to produce ordinary interference fringes.

Let us go back to the analogy that speckle rays behave as if they were reflections from small mirrors that are fixed to the object. (For the limitations of this analogy see Section 4.2.8). In Fig. 5.5(a), (b) and (c) three different object motions are seen with the accompanying motions of one speckle ray. A is the point of illumination at a distance L from the object; B is the point of observation; C is the scattering point at the object. The interference fringes seen from B appear to be localized at the point Q, where the homologous rays C_1B_1 and C_2B_2 intersect.

In Fig. 5.5(a) the object is translated vertically from C_1 to C_2 and the rays reflected from the same small mirror will be reflected as if they came from a point at a distance L behind the object. Thus the localization of the fringes will be at the same distance behind the object as the illumination source is in front of it.

In Fig. 5.5(b) the object is rotated through a very small angle around an axis at a distance R behind its surface. It is seen that the homologous rays intersect half-way towards the axis of rotation (provided that L is large compared with R and that the studied surface is normal to the directions of illumination and observation). Thus the interference fringes will appear at a distance $0.5R$ behind the surface.

Finally, it is seen from Fig. 5.5(c) that if the axis of rotation is in front of the object then in this case the fringe localization will also be approximately half-way towards the axis of rotation. Thus the interference fringes will appear at a distance $0.5R$ in front of the surface.

5.3.3 The focusing effect from "motion trails"

In this section a new analogy to fringe localization is introduced which to the knowledge of the present author has not been published before. Its merits are that it is so simple that one quickly gets an intuitive feeling for the fringe localization and also that it produces the correct answer even for the most extreme cases, when other methods become very complicated.

When studying Fig. 5.5(a), (b) and (c) the impression is gained that what is seen is simply the focusing effect of mirrors. In the case shown in Fig. 5.5(a) the observer sees the illumination point A at the distance L behind the object, as if reflected by a flat mirror between C_1 and C_2. (The motion was along a straight line.) In Fig. 5.5(b) the observer sees the illumination point A at the distance $0.5R$ behind the object, as if reflected by a convex mirror. (The motion was along a convex curve.) In Fig. 5.5(c) the observer sees the illumination point A at the distance $0.5R$ in front of the object, as if reflected by a concave mirror. (The motion was along a concave curve.)

If it is also included in the calculation that L is *not* infinitely larger than R, then it is obvious that in Fig. 5.5(c), for example, the intersection of the homologous rays will be projected further out in front of the object. This will also be the case with the image reflected by the mirror C_1C_2.

To study the focusing effect more closely, consider Fig. 5.6. A small mirror on the surface of the object is first positioned at C_1. Between the two exposures the object is rotated through the angle β around the point O at the distance R so that the mirror moves a distance h to C_2. The illumination is made from a point at a distance L. The point of intersection of the homologous rays is calculated in the following way (Fig. 5.6):

$$\gamma = (\beta - \alpha) + (\beta - \alpha) + \alpha, \tag{5.3}$$

but for small values of α, β and γ,

$$\alpha = h/L, \tag{5.4}$$

$$\beta = h/R, \tag{5.5}$$

$$\gamma = h/X. \tag{5.6}$$

Thus

$$\frac{1}{X} = \frac{2}{R} - \frac{1}{L} \tag{5.7}$$

or, in the more usual form for an image focused by a concave mirror,

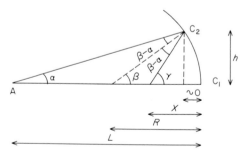

Fig. 5.6. It is deduced that the concept of homologous rays (Fig. 5.5) produces a result that is identical to the new concept that the motion of a scattering point at the object surface produces a reflecting trail. Light from A at a distance L is reflected by the object point which, because of a rotation of the radius (R) through the angle β, has moved from C_1 to C_2. The localization is at a distance X in front of the object.

$$\frac{1}{F} = \frac{1}{X} + \frac{1}{L},\qquad\qquad(5.8)$$

where F is the focal length and is given by $F = R/2$.

From equation (5.8) it is evident that a calculation using the concept of homologous rays produces a result that is identical to the new concept that the motion of a scattering point at the surface of the object produces a reflecting trail. For these curved reflecting trails the term "motion trails" will be introduced. The apparent localization of the holographic fringes is thus explained as being caused by the integrated focusing effect of a great number of scattering motion trails that work together like an incoherent grating.

For an alternative deduction of the above statement it is possible to go back and consider the ellipses of the holo-diagram. One definition of fringe localization could be that if the point of observation is placed where the fringes are localized, then the object appears to be fringe-free, being either bright or dark. It is analogous to the case of a fence with vertical bars: when an observer walks up to such a fence and looks through it at zero distance, he sees no bars!

From the definition of the holo-diagram no fringes are formed when the motion is along the surface of one ellipsoid. Thus an observer can move his eyes until he finds a position from which he sees no fringes. He then concludes that his eye is in the plane of localization of the fringes. He also knows that it is in one focal point of the ellipsoids, the other one being the point of illumination. Thus it has again been proved that the fringes localize where the light would be focused by mirrors having the shape of the curve of motion of each point on the object.

Using this concept of scattering motion trails it can be understood that most motions produce no true imaging effect and thus no true localization. Rigid body translation results in flat mirrors which produce a perfect image and therefore a perfect localization. Rigid body rotation, as already described, produces mirrors with circular cross-sections which produce a good imaging effect without aberrations only for a small area. Thus the described focusing effect is, in that case, only an approximation and different observation angles produce different localizations of the fringes. Perfect localization is produced only by motions along paths that form straight lines, ellipses, parabolas or hyperbolas.

5.3.4 Extreme examples

To see if these results agree with reality, consider a motion that is exclusively along the ellipsoids of which one focal point is the point of illumination. Then all the fringes should form at a single point, the other focus of the ellipsoids. It is, however, very difficult to deform an object in such a way that each surface point moves along an ellipsoid. Let us therefore reconsider the experiment already described in Section 4.4.4 together with Figs 4.23, 4.24 and 4.25.

A diffusely scattering ground glass was placed at C while it was illuminated from A (Fig. 4.23). A hologram plate (H) was placed closer to C than A and at the opposite side. The ground glass was displaced a distance d between the two exposures of the plate (H) which was also illuminated by a reference beam (R). When the reconstruction was studied the ground glass (C) was seen covered by fringes that became circular when the observation was made from the axis AB (see Fig. 4.25).

The circular fringes became broader and their number decreased as the observation point was moved away from the hologram plate towards B. When the eye of the observer was exactly at B (so that AC = CB) the fringes disappeared completely; as B was passed the circular fringes appeared again and grew in number. When the observation was made from beyond B, the circular fringes moved to the right as the eye was moved to the left. When the observation point was exactly at B the parallax effect became infinite and as B was passed it changed sign. Finally, it was also found that circular fringes were focused at a ground glass placed at B during the reconstruction of the virtual image of C.

Thus all the experimental facts indicate that the fringes were localized at B, which was at the same distance (l) from the displaced screen (C) as the point of illumination (A) during recording.

Now let us study the focusing effect of the scattering motion trails. When the ground glass (C) was displaced by a distance d, each scattering surface point moved along a line parallel to AB. The straight motion trails work like mirrors that focus the light from A to B by reflection, as demonstrated in Fig. 5.7. This simple figure also explains why circular fringes are formed when the eye is in front of, or behind, the point B. Furthermore, it explains the parallax effect completely. The analogy between Fig. 5.7 and the holographic experiment is not simply an approximation: the location of the centre of the circular fringes and also the movements of the fringes are identical for the two cases. The only difference is that just a single fringe is seen in the motion trail analogy.

We have now studied one example of out-of-plane translation. The case of out-of-plane rotation, which means rotation round an axis in the plane of the surface of the object, has already been covered by Fig. 5.5(b) and (c). In-plane motion in a straight line was studied in Fig. 5.5(a). What remains is to use the fringe location analogy to study one of the most complex problems, that of fringes caused by a rotation around an axis that is normal to the surface of the object. The points on the surface will move along the circumferences of circles, so it is possible to study the light scattered by a gramophone record or any other circular disc that has been turned or ground

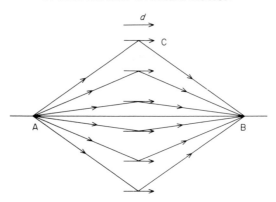

Fig. 5.7. The ground glass (C) of Fig. 4.23 was illuminated from A. Between the two exposures of the hologram it was displaced by translation in the direction *d*. The "motion trails" represented by the arrows are all parallel to *d*. The reflections from these motion trails are focused to a point B which will be the localization of the fringes. This result is confirmed by Figs 4.24 and 4.25.

or machined in such a way that it has got circular grooves around its centre. These grooves will scatter light in the same way as the motion trails produced by the circular motion of all the scattering points on a disc given a small rotation around its centre between the two exposures of a hologram.

Place the disc flat on a table and illuminate it with light from a single point-like source placed some distance away. One bright straight fringe will be seen that always passes through the centre of the disc. In whatever way the light source is moved, this diametral fringe will constantly bisect the angle separating the directions of illumination and observation, respectively. This fringe, caused by the directional scattering effect of the grooves, is identical to the zeroth-order fringe seen on the holographic reconstruction of a diffusely scattering disc that is slightly rotated around its centre between the two exposures of a hologram. The only difference is that in holographic interferometry there are higher-order fringes more or less parallel to the zeroth-order fringe. Otherwise the direction in relation to illumination and observation will be identical. The fringe on the disc will also move, as the point of observation is changed, in exactly the same was as the corresponding interference fringes. Thus even the apparent localization in space will be an exact copy of that of the zeroth-order interference fringe. When the observer moves his head from left to right the fringe reflected by a concave groove will appear to be localized in front of the disc while that reflected by a convex groove appears to be localized behind the disc.

The analogy of reflecting motion trails is especially useful in those extreme cases when the mathematics becomes so complex that no solutions have, as yet, been published. Thus, if for example the observer brings his eye closer

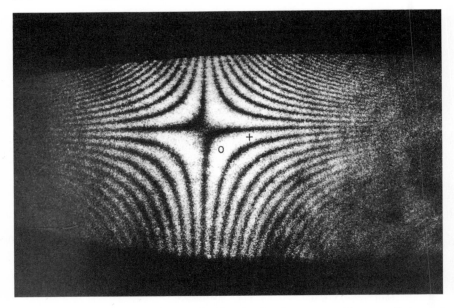

Fig. 5.8. The reconstructed image of a disc that has been rotated by a small amount between the two exposures of a hologram. The disc is seen from a very oblique angle. The centre of rotation is marked by a cross (+). The point where one of the ellipsoids of the holo-diagram is tangential to the disc surface is marked by a small circle (○).

and closer to the disc he will finally see how the originally straight bright fringe becomes curved and finally breaks up into the two branches of a hyperbola. The behaviour of the fringe pattern is exactly the same as that which will be explained by the use of the holo-diagram in Section 5.4.5 (see Figs 5.12, 5.13) and also by the moiré pattern analogy of Section 5.5.5 (see Fig. 5.18).

A photograph of a disc rotated between the two exposures is seen in Fig. 5.8, while the corresponding fringes from a disc with circular grooves is seen in Fig. 5.9. The optical configurations used for making Figs 5.8 and 5.9 were identical, the disc was so positioned that it was tangential to one of the ellipsoids of the holo-diagram. The hologram plate was so close that it almost touched the rotated disc. In whatever way the observation point was moved, the centre of rotation remained in one of the branches of the hyperbola, while the tangent point remained in the other. To produce the fringes of Fig. 5.9 it is better not to use a gramophone record but rather some other disc that has got circular grooves all the way to its centre. The fringes at the central cross-over point of Fig. 5.9 break up and recombine when the angle of observation is changed. This behaviour cannot be explained as a parallax effect caused by a fixed fringe localization in space.

Fig. 5.9. A steel disc has been turned in a lathe so that its flat surface is covered by circular grooves. These grooves work like an incoherent grating and reflect light as in the "motion trails" of Fig. 5.8. Thus the two methods produce identical patterns when illuminated and observed in identical configurations. The difference is that holographic interferometry produces a set of fringes while the analogy based on reflections from "motion trails" only mimics the zeroth-order fringe.

5.4 Fringe pattern on the object

We have studied how the area of the hologram plate can be utilized for evaluation of the fringes caused by in-plane motion. This was done by moving the eye behind the plate while observing one single point at the object. Let us now make the evaluation the other way round, i.e. study the whole object surface from one point of the hologram plate.

If there was no extreme point at the object surface, there would be no fringes. Let the object be flat, the motion be a translation without rotation, and the illumination and the observation be made from infinite distances. In that case no fringes will accumulate on the object surface because every point will be equal to the others. The only effect will be that if the object is studied in real-time holography its total surface will alternate between bright and dark during the movement. However, when the illumination or the observation is carried out from a short distance, or if the displacement varies over the object surface, then fringes will accumulate. The reason for this is that there

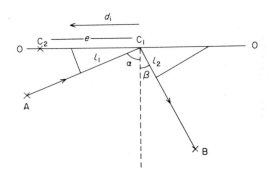

Fig. 5.10. Calculation of the path length difference caused by the in-plane motion d_i. OO is the object surface with the point C_1 under study. A is the illumination source and B is the point of observation.

will, during the motion, be a beating effect in the light intensity, having different rates at different points on the surface.

Let us start by using simple mathematical calculations (see Fig. 5.10) to see how the path length difference varies over a flat surface (OO) that is translated in the vicinity of the point of illumination (A) and the point of observation (B). The point (C_1) on the object is translated in-plane by a distance d_i. The illumination distance is R while the observation distance is L. Let us calculate the distance (e) between two adjacent fringes formed on the object surface.

The displacement d_i results in a decrease (l_1) of the path length of AC_1 combined with an increase (l_2) of the path length of $C_1 B$:

$$l_1 = d_i \sin \alpha, \tag{5.9}$$

$$l_2 = d_i \sin \beta. \tag{5.10}$$

The total path length difference is $l_1 - l_2$, resulting in n cycles of brightness and darkness. Thus,

$$d_i(\sin \alpha - \sin \beta) = n\lambda. \tag{5.11}$$

Considering small increments $\delta\alpha$, $\delta\beta$ and δn in α, β and n, respectively, it can be shown that

$$d_i(\delta\alpha \cos \alpha - \delta\beta \cos \beta) = \delta n\lambda. \tag{5.12}$$

Let us now see how far to the left (Fig. 5.10) another point C_2 has to be located so that just one fringe is found in the distance e between C_1 and C_2. Thus,

$$\delta n = 1, \tag{5.13}$$

$$\delta\alpha = \frac{e \cos \alpha}{R}, \tag{5.14}$$

$$\delta\beta = -\frac{e\cos\beta}{L}. \tag{5.15}$$

By inserting equations (5.13), (5.14) and (5.15) into equation (5.12) it is found that

$$e = \frac{RL}{d_i}\frac{\lambda}{L\cos^2\alpha + R\cos^2\beta}, \tag{5.16}$$

where e is the fringe separation at the object, R is the distance to the source of illumination, L is the distance to the point of observation, d_i is the in-plane motion, α is the angle between the plane of illumination and the normal to the surface, and β is the angle between the plane of observation and the normal to the surface.

5.4.1 Young's fringes of observation

Let us now make an identical calculation but this time use the concept of an analogous grating at the object (Section 4.2.8) combined with Young's fringes of observation (Section 4.2.7). This means that it is necessary to take into consideration how the speckles move on the hologram plate (see Fig. 5.11).

Start with the grating equation (4.17):

$$g(\sin\alpha - \sin\beta) = \lambda.$$

Considering small increments in α and β it can be shown that

$$\delta\beta = \delta\alpha\frac{\cos\alpha}{\cos\beta}. \tag{5.17}$$

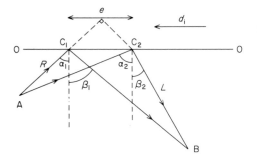

Fig. 5.11. Based on Fig. 5.10 a study is made of the separation (e) of the fringes caused by the in-plane displacement (d_i) if R is the distance to the illumination source A, and L is the distance to the point of observation B. C_1 and C_2 are two points on the surface OO separated by one fringe.

As the in-plane displacement (d_i) is infinitesimal compared to R,

$$\delta\alpha = \frac{d_i \cos \alpha}{R}. \tag{5.18}$$

Thus,

$$\delta\beta = \frac{d_i \cos^2 \alpha}{R \cos \beta}. \tag{5.19}$$

The speckle motion (h) on the hologram plate at the distance L is therefore given by

$$h = \delta\beta L + d_i \cos \beta. \tag{5.20}$$

The twin speckles at the plate separated by h would, in the true real image (Section 2.3.6), project the Young's fringes with the separation f:

$$f = \frac{\lambda}{2 \sin \gamma} = \frac{\lambda L}{h} = \frac{\lambda L R \cos \beta}{L \cos^2 \alpha + R \cos^2 \beta}. \tag{5.21}$$

Finally, because the normal to the object's surface is inclined at an angle β to the projection of the Young's fringes, the fringe separation distance (e) at the reconstructed object surface will be given by

$$e = \frac{RL}{d_i} \frac{\lambda}{L \cos^2 \alpha + R \cos^2 \beta}. \tag{5.22}$$

The true real image reconstructed by a beam that is antiparallel to the reference beam is identical to the true virtual image, so therefore the same fringe separation is also found in the virtual image as projected by Young's fringes of observation.

Equation (5.22) is identical to equation (5.16), which was based solely on calculations of path lengths. Therefore it is concluded that the analogous grating and also the concept of Young's fringes of observation are valid for in-plane motions as well as for tilt (Section 4.2.8). From equations (5.22) and (4.20) it can also be understood how a manipulation of the separation of the twin speckles, e.g. by a shearing or a tilt of a sandwich hologram, will influence the fringes. Thus a tilt of the sandwich hologram so that h of equation (5.20) becomes zero will compensate for the fringes caused by in-plane motion just as well as the fringes caused by tilt are compensated for by eliminating h of equation (4.20).

5.4.2 Approximate methods

Let us return to equation (5.16), which is fundamental to the fringe formation caused by in-plane motion. If the illumination is collimated $(R = \infty)$ and if the direction of illumination is perpendicular to the object's surface ($\beta = 0$) then the equation can be simplified:

$$e = L\lambda/d_i. \tag{5.23}$$

This equation indicates that the fringe pattern seen on the object is equivalent to the Young's fringes of observation that would have formed if the object had been fixed and, instead, the point of observation had been displaced in a direction and magnitude that were identical to that of the object.

Had the illumination not been collimated, but rather emitted from a point close to the point of observation, the above statement would still be true, except that the sensitivity would be twice as high. In that case equation (5.16) is transformed into

$$e = L\lambda/2d_i. \tag{5.24}$$

The fringes on the object surface have been described as Young's fringes projected from the point of observation, which is thought of as being given a small displacement identical to that of the displacement of the object. This analogous method is useful for a quick intuitive understanding of the fringe patterns caused by all kinds of rigid body motion. It must, however, be kept in mind that it is an approximation limited to illumination and observation in a direction that is normal to the surface under examination. It is also limited to such configurations that the distance between A and B on the holo-diagram is small compared to the distance to the object.

What will the error be if the simplified equation (5.24) is used instead of the correct equation (5.16)? Let R be equal to L and let the distance AB on the holo-diagram be $0.5L$ instead of the value of zero which is needed for equation (5.24) to be correct. The result will be only 6.5% too small, an error that can very easily be accepted.

Thus it has been found that when the object is at large distance any in-plane displacement will cause Young's fringes of observation which will be perpendicular to the direction of displacement. As described in Section 4.4.3, the Young's fringes analogy is also valid for out-of-plane translations which cause a set of circular fringes.

Fringes caused by rotation around an axis that is normal to the surface can, however, not be explained by a Young's fringes analogy because a rotation of a point source causes no fringes. Other methods have to be used for this case and for the evaluation of fringes formed on objects that are in the near field of the holo-diagram, which was defined in Section 4.4.2 as the area within a circle with radius AB and its centre half-way between A and B ($L = $ AB). Equation (5.16) *can* be used, but it is rather complicated in spite of being only two-dimensional; also, as a consequence, it can be used only to calculate distances along a line, not to explain the curvature of fringes covering a surface.

5.4.3 The holo-diagram for translation

Section 4.4 described how the holo-diagram can be used to evaluate fringes caused by out-of-plane motion. The diagram can just as well be used as a practical device to understand the fringe patterns formed by rigid in-plane motions.[6]

Starting with translations, equation (4.39) and Fig. 4.22 describe the fringe pattern formed on a flat surface positioned infinitely close "behind" the points A and B and translated in the x-direction between the two exposures. The k_x contours represent the loci of constant sensitivity and thus they also represent the interference fringes caused by a translation that is constant over the whole surface. If the surface of the object under study had been far behind the points A and B, the fringes would have been vertical, straight and equidistant. In the near field of AB, however, this pattern is transformed into that of Fig. 4.22. The reason for this transformation is that there exist two areas of low sensitivity. One area is between A and B where the k-value of the holo-diagram increases to an infinite value. The other area is along the y-axis where the sensitivity is low because the ellipses are parallel to the displacement in the x-direction.

A translation in the y-direction of a flat surface placed infinitely close behind A and B causes a fringe pattern explained by equation (4.43) and Fig. 4.26. The k_y contours represent constant sensitivity to translations in the y-direction, but at the same time represent fringes caused by such movements. If the surface under study had been far behind the points A and B the fringes would have been horizontal, straight and equidistant.

In the near field of AB this pattern breaks down to that of Fig. 4.26. The reason for the rather complicated pattern close to AB is that there exists one area of low sensitivity between A and B where the k-value of the holo-diagram increases to an infinite value. There also exist areas of zero sensitivity at the x-axis outside A and B because at these areas the motion in the y-direction is parallel to the ellipses. As described in Fig. 4.26, at a certain distance from AB there is an "island" of almost constant sensitivity (straight k_y contours).

5.4.4 Rotation in the plane of the holo-diagram

It is, for practical reasons, impossible to study a surface in the plane of the diagram because in that case both illumination and observation are parallel to the disc's surface. Let us therefore study a disc positioned at such a small distance beneath the (horizontal) plane of the holo-diagram that the influence of this distance on the fringe pattern can be neglected. The disc surface is parallel to the plane of the diagram and its axis is normal to that plane.

As the disc in Fig. 5.12(a) rotates it has two lines of zero sensitivity to fringe

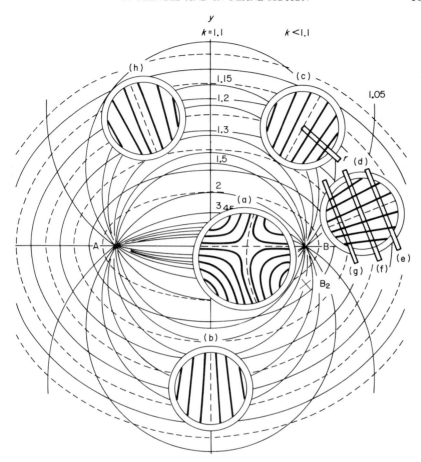

Fig. 5.12. Interference patterns formed on the surfaces of discs that have been given a small rotation between the two exposures of double-exposure holograms. The discs are supposed to be placed just behind the surface of the diagram.

formation (broken lines in the figure). One line goes along the x-axis, where the k-value is infinite. Another line (which is slightly curved) goes along all those points where the movement of the disc surface is parallel to the ellipses. This latter line passes through the centre of the disc. Between these two lines that intersect at the x-axis there exist two lines of maximal sensitivity that also intersect at that point. Therefore, the fringes must form a set of hyperbola-like lines that at no time cross the two lines of zero sensitivity. A person who studies this fringe pattern will, judging from the parallax effect, say that the fringes are localized at a distance far behind the surface, and if A is at infinite distance the fringes also appear to be at infinite distance. This effect occurs because the localization of the asymptote that coincides with the line AB is dependent only on the position of the point of observation and is not influenced by the position of the centre of the disc.

If the area of the disc covers the x-axis, the pattern will be very similar to that formed if the centre is at the x-axis. If the radius is so small that the disc surface does not reach the x-axis, there will exist only one line of zero sensitivity, formed by those points on the disc surface that move parallel to the ellipses (Fig. 5.12(b)).

The component of the disc motion that is normal to the ellipses (nearly vertical in the figure) increases in proportion to the (horizontal) distance from the centre of the disc. Therefore, the disc will be covered by fringes that are more or less parallel to the line of zero sensitivity which is at no time crossed by any fringe. These fringes bisect the angle between illumination and observation.

Thus the bright fringe nearest to the centre (double-exposure hologram) seems to be fixed to the disc while the other fringes move in relation to the disc's surface as the point of observation is changed: the further away from the centre the greater is this movement. This effect can give the impression that the fringes are fixed to a surface that is inclined in relation to the disc surface and intersects this surface at the centre of the disc. However, if the point of observation is changed too much, this impression is destroyed because the fringe pattern is changed in such a way that it cannot be explained by any fixed localization in the depth of the fringes.

A bar is placed on top of the disc (Fig. 5.12(d)/(e)) and rotates with the disc. An observer at B will see fringes on the side of this bar. A movement of B downwards (to B_2) will result in an upward movement of the fringes on the bar. (The fringe through the centre of the disc pivots around this centre because it is not allowed to cross it, and all fringes are, at this part of the diagram, more or less directed towards B.) Therefore, the fringes on the side of the bar that is positioned as shown in Fig. 5.12(d)/(e) seem to be localized in front of its surface. In the case of the bar positioned as shown in Fig. 5.12(d)/(f) the fringes seem to be more or less localized at the surface, and in the case of the bar positioned as shown in Fig. 5.12(d)/(g) they seem to be behind the surface.

This result could be compared to that of Fig. 5.5(c). It also could be confirmed by a simple experiment using a gramophone record and the analogy of reflection from motion trails.

5.4.5 Rotation in space

Thus far we have studied a disc rotating in the plane of the holo-diagram and found that it is simple to predict the fringe pattern in a geometrical way. The situation changes little as the disc moves away from the plane of the diagram as long as we are satisfied with the approximate appearance of the fringe pattern. The holo-diagram represents only one cross-section of the ellipsoids; other cross-sections can be studied in exactly the same way. The simplest method to use for the prediction of the fringe pattern is to study the lines formed by the intersections of the ellipsoids and the plane surface of the disc. These intersections form a set of concentric ellipses. The principal difference from the ellipses in the holo-diagram is that they do not have common focal

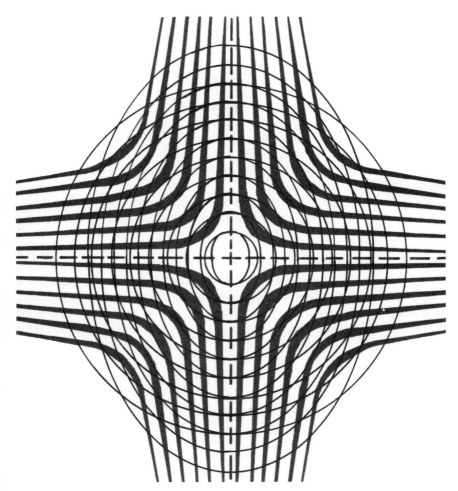

Fig. 5.13. Fringe pattern at the point where a flat surface is tangential to one ellipsoid. The flat surface is rotated around the tangential point. Two (broken) lines of zero sensitivity are formed where the circles representing direction of displacement are parallel to the ellipses, which are cross-sections of the set of ellipsoids forming the holo-diagram.

points, but the distance between those points decreases as the size of the ellipses decreases. One way to obtain a simple view of the cross-sections is to look at the cross-section of an onion that has been cut through with a sharp knife.

If the disc is placed with its axis coincident with AB, the intersections of the ellipsoids will form circles on the disc surface. When the disc is rotated around the axis AB, no point on its surface will ever cross any circle, and therefore the whole area of the disc has zero sensitivity to fringe formation. This is also the case if a disc with limited area is positioned anywhere at an infinite distance from AB and is rotated round an axis through the point where it touches one of the ellipsoids (which at that distance are transformed into spheres).

A plane opaque surface placed in the vicinity of AB so that it is illuminated from A and observed from B will always have a tangent point to one of the ellipsoids. At the tangent point, the normal to the surface of the disc coincides with the normal to the ellipsoids with which it makes contact. As the disc rotates around this axis, each point on its surface will cross the same ellipse four times and also move parallel to that ellipse four times. Therefore, two lines of zero sensitivity are formed which intersect at the centre of the disc, as demonstrated in Fig. 5.13.

The easiest way to predict the fringe pattern formed on the surface of a disc rotating round an axis that is normal to its surface, and in the vicinity of the point where the disc just touches one of the ellipsoids, is as follows.

Draw a set of concentric ellipses representing the intersections of the ellipsoids and the disc surface (Figs 5.13 and 5.14). Around the axis at the centre of the disc is a set of concentric circles, also drawn to represent the direction of motion of points on the disc surface. Where ellipses and circles are parallel, the sensitivity to fringe formation is zero. There are two lines of zero sensitivity (broken lines in the figures), one of which passes through both the centre of the disc and the tangent point.

5.4.6 Fringe localization in space

A vertical movement of the eye studying the hologram in Fig. 5.14 will give the impression that the fringes are localized at the surface of the disc because the line of zero sensitivity is fixed at the centre of the disc. A horizontal movement of the eye, however, will give the impression that the fringes are localized far behind the surface of the disc because the line of zero sensitivity is also fixed at the centre of the tangent point between disc and ellipsoids. This point moves in the same direction relative to the surface of the disc as the eye, which gives the impression that it might be localized somewhere behind the disc.

The pattern in Fig. 5.14 was photographed using a camera with an

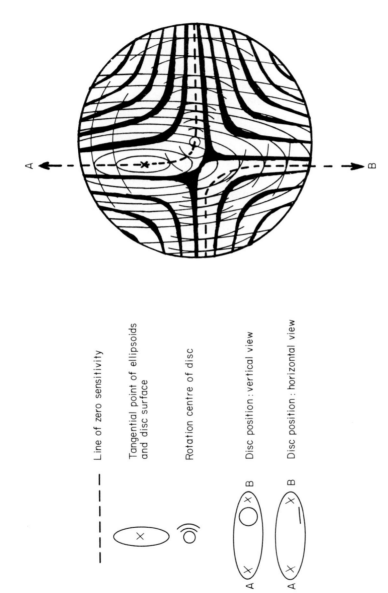

Line of zero sensitivity

Tangential point of ellipsoids and disc surface

Rotation centre of disc

Disc position : vertical view

Disc position : horizontal view

Fig. 5.14. Same as Fig. 5.13 but the centre of rotation does not coincide with the tangent point. The two broken lines of zero sensitivity do not intersect.

aperture in the form of a narrow slit. The camera had to be focused at a distance of less than 1 m when the slit was vertical. With a horizontal slit, the maximal resolution of the fringes was produced when the camera was focused at infinite distance (Fig. 5.8).

If the centre of the disc is far to the right or left in Fig. 5.13 or Fig. 5.14, and if the diameter of the disc is so small that the tangent point is not reached by the surface of the disc, only a set of horizontal fringes are seen. These fringes can appear to be fixed to a surface that is inclined with respect to the surface of the disc and intersects this where the bright fringe through the disc's centre is formed. This fringe represents the line of zero sensitivity and therefore has to be fixed to the centre of the disc regardless of the way in which the observer moves his eyes. If the centre of the disc is moved far upwards or downwards in Fig. 5.13 or Fig. 5.14, the results is similar, and a set of vertical fringes are formed. If the disc's centre is moved to an infinite distance from the tangent point, the motion is transformed into pure translation.

Almost everything that has been described here about the holographic interference fringes formed on a rotated disc and their parallax effect can be verified by the reader using the analogy of the gramophone record illustrated in Fig. 5.9. This method is not merely an analogy based on approximations; the basic equations are identical for the two phenomena. One great limitation, however, is that the gramophone record will produce only the zeroth-order fringe while measuring techniques based on holographic interferometry need the whole set of fringes. Thus in the next section the moiré pattern analogy to hologram interferometry will be examined; this is another analogical method of evaluation that includes the total fringe pattern.

5.5 The moiré pattern analogy to hologram interferometry

The formula for the interference fringes formed on an object that has been given a small in-plane displacement between the two exposures of a hologram was found earlier (equation (5.16)). This equation describes how the fringe spacing depends on the displacement and the angles of illumination and observation. However, it only gives this function along one line. To find the fringe pattern formed on a surface in the near field of the holo-diagram the theorem of Pythagoras was used to construct the equations (4.39) and (4.43) on which Figs 4.22 and 4.26 are based.

These equations are rather complicated and each still represents just one single direction of motion at a single distance (zero) from the axis of the holo-diagram. The visualization of all the patterns formed by any random in-plane translation of a three-dimensional object would appear to be an almost impossible task, at least if the method is to be quick, easy to use and, last but not least, easy to understand. However, the moiré pattern analogy to hologram interferometry comes very close to this goal.

The best way to explain the moiré pattern analogy to hologram interferometry is probably as follows:

(1) To be observed an object must scatter light. If it is illuminated from a point source A and observed from a point B, only those scattering points are used for the observation which represent a grating that everywhere has such a frequency that it diffracts light from A to B.

(2) If the object is displaced (e.g. tilted) between the two exposures of a hologram, then the grating frequency utilized must be different for the two situations.

(3) The number of holographic interference fringes seen on an object which has been moved is equal to the difference in the number of grating lines (grooves) needed before and after movement, respectively, to diffract light from the light source to the observation point.

(4) The simplest way to find the difference in number of grooves between the two gratings is to put one on top of the other. The number of moiré fringes resulting will be equal to this number.

(5) The grating needed on the surface of any three-dimensional object to diffract light from A to B is represented by the intersections of the object's surface with the ellipsoids of the holo-diagram (Section 1.4.3)

Thus the fringes formed in holographic interferometry can be evaluated by studying the moiré effect of the intersections between the object's surface and the ellipsoids during the first and the second exposures respectively.

The most crucial statements (numbers 2 and 3) have already been derived mathematically in Section 4.2.8, resulting in equation (4.23). Statements 1 and 5 are accepted as results from the definition of the holo-diagram. Statement 3 has, however, been proved for a tilted object only; it now remains to provide a similar proof for translated objects.

5.5.1 Mathematical verification

In the following it will be proved that the moiré pattern analogy to holographic interferometry is valid not only for tilt (equation (4.23)) but also for translation (Fig. 5.11). The length of the object (C_1C_2) is e and it is displaced a distance d while illuminated from A at the angle α and observed from B at the angle β. Let us start by calculating the number of fringes formed between C_1 and C_2 using only the differences in path lengths.

The number of fringes (n_1) passing C_1 during the displacement (d) will be

$$n_1 = \frac{d(\sin \alpha_1 - \sin \beta_1)}{\lambda}. \tag{5.25}$$

The corresponding number of fringes (n_2) passing C_2 will be

$$n_2 = \frac{d(\sin \alpha_2 - \sin \beta_2)}{\lambda}. \tag{5.26}$$

The number of fringes accumulated between C_1 and C_2 due to the displacement d will be $n_1 - n_2$. Thus,

$$n_1 - n_2 = (d/\lambda)(\sin \alpha_1 - \sin \beta_1 - \sin \alpha_2 + \sin \beta_2). \tag{5.27}$$

Now make the corresponding calculation using the grating analogy of Section 4.2.8.

A grating at C_1 that diffracts light from A to B will have a separation g_1 of the grating lines (equation (4.17)):

$$g_1 = \frac{\lambda}{\sin \alpha_1 - \sin \beta_1}. \tag{5.28}$$

Similarly, at C_2,

$$g_2 = \frac{\lambda}{\sin \alpha_2 - \sin \beta_2}. \tag{5.29}$$

Between C_1 and C_2 the value of g will vary in an unknown way. Let us, however, make a copy of this grating and put it on top of the original one. No moiré fringes are formed at first because the two gratings are identical, but if one grating is translated with respect to the other moiré fringes will form that represent the difference in grating frequencies (lines per unit length).

When the object is displaced by d there will be n_1 lines of the top grating which move outside the lower grating at C_1, while n_2 lines of the lower grating move outside the top grating at C_2. Thus the difference $(n_1 - n_2)$ will also represent the difference in the number of grating lines on the remaining parts of the gratings that cover each other. Consequently, $n_1 - n_2$ will be the number of moiré fringes formed:

$$n_1 = d/g_1, \tag{5.30}$$

$$n_2 = d/g_2, \tag{5.31}$$

$$n_1 - n_2 = (d/\lambda)(\sin \alpha_1 - \sin \beta_1 - \sin \alpha_2 + \sin \beta_2). \tag{5.32}$$

In this way we have verified that for in-plane displacements the number of moiré fringes formed by analogous gratings (5.32) is equal to the number of holographic interference fringes as calculated in the conventional way solely from the difference in path lengths. This identity proves that the methods of Fig. 3.9 and Fig. 3.24 represent not only approximate visualizations but are mathematically correct analogies. The figures, however, represent only the fringe-forming process when illumination and observation are made from infinite distances. When this process has to be studied in the near field the flat, equidistant, imaginary interference surfaces have to be replaced by the ellipsoids of the holo-diagram.

5.5.2 The moiré fringe-forming process at the object

Any two sets of fringes that when combined produce a new set of fringes are known as primary fringes, and the new fringes are called secondary fringes. A combination of two sets of secondary fringes can also produce difference fringes of the third order (tertiary fringes), etc. The higher-order fringes represent the difference of two lower-order fringe patterns.

Thus, the term primary fringes refers, for example, to those fringes on a hologram plate that produce the holographic image. The dark fringes in hologram interferometry that represent displacement, deformation, vibration, or changes in refractive index all are secondary fringes.

Ordinary moiré fringes are secondary fringes, e.g. produced by placing two transparencies covered by primary fringes on top of one another. As both moiré patterns and hologram interference patterns are secondary fringes, there might be reasons to believe that useful analogies exist between these two phenomena. This fact was also verified mathematically in Section 5.5.1.

Let us repeat the construction of a double-exposure hologram. The object C is illuminated by a spherical coherent wavefront originating from a point source A. The observation is made from another point B at a distance L from A. A photographic plate P is placed between C and B, the purpose of which is to memorize all the optical information from C. (To accomplish this, P also must be illuminated by a reference beam.)

The light is turned on for the first exposure and then turned off. Thereafter the object is deformed so that parts of it are displaced while other parts remain in a fixed position. When the object has settled down, a second exposure is made, and thereafter the photographic plate is processed and brought back to its original position where it is again illuminated by the reference beam. If the object C is taken away, its reconstructed holographic image will be seen at exactly the place where it was during exposure. When the image of the object is studied from the observation point B, interference fringes are found, fixed to its surface, that represent the displacements of each surface point. The fringe pattern can be interpreted in the following way.

Let A and B represent the common focal points of a set of rotationally symmetric ellipsoids, each one representing a constant path length for the light transmitted from A to B via any particular point on the ellipsoidal surface (Fig. 1.24(a)). Let the difference in path length be half of the wavelength (λ) if the reflection is changed from one ellipsoidal shell to the adjacent one. Number the ellipsoidal shells $s_1, s_2, \ldots, s_n$ and the corresponding path lengths are $L + \lambda/2, L + 2(\lambda/2), \ldots, L + n(\lambda/2)$.

Any point C on the surface of the object that, between the two exposures, has moved from one ellipsoid (s_k) to another (s_{k+q}), where q is an odd number, will look dark in the reconstruction. A point that has moved so that q is zero or an even number will cause a bright reconstruction. Thus $q/2$ dark fringes are formed on the surface of the object between the displaced point C and a fixed point (if no extreme point exists in between).

The moiré pattern equivalent to the described fringe formation is quite simple in theory.[7,8] Fill every second space between the adjacent imaginary ellipsoids with an imaginary ink that is in an inactive state. Place the object in the space occupied by the set of ellipsoids that it intersects. Activate the ink so that it marks the object's surface black on all those parts that are in contact with the imaginary ink. This process corresponds to the first exposure of the hologram plate. The object is now covered by black fringes that are so closely spaced that they cannot be seen by the naked eye. These fringes are named primary fringes.

Deform the object and once more activate the ink. A new set of primary fringes is formed on the object surface. This process corresponds to the second exposure of the hologram plate. If the displacement between the two exposures is only a few wavelengths, the two sets of primary fringes will result in a moiré effect, and a new fringe pattern is formed that can be seen fixed to the surface of the object. These moiré fringes, which are called secondary fringes, represent the displacement of each surface point and are identical in number and shape to the holographic interference fringes.

This analogy is true regardless of the type of motion (translation, deformation, or extension). It is also independent of the type of holographic set-up or the shape of the object. The only restriction on the use of the moiré pattern method is that illumination and observation have to be made from apertures that are small compared to the distances to the object. However, this restriction also applies to holographic interference fringes. *The larger the apertures of illumination and observation the less distinct are the interference fringes.*

The equivalent process to real-time observation of holographic fringes (live fringes) is the following. The imaginary ink should be transparent, but even in its inactive state it should look black at those parts where it contacts the object surface. The exposure of the hologram plate corresponds to activation of the ink, so that the object becomes covered by primary fringes fixed to its surface. Thereafter the ink is made inactive again, and a second set of primary fringes is formed on the object on all those parts where its surface is in contact with the ink. The fixed and "live" sets of primary fringes form secondary fringes that are analogous to the "live" interference fringes that are seen in real-time holography.

If this analogy is to be absolutely strict, then between the first exposure and the observation the ink should be moved from the filled spaces between the ellipsoidal shells into those that earlier were empty. This process corresponds to the development of ordinary photographic material in which illuminated areas are darkened.

When an interference hologram is studied, the interference fringes move around as the observer moves his point of observation (the focal point B). Consequently, if that point of observation is to be changed, the whole process of making the two sets of primary fringes has to be repeated. An instrument has, however, been presented that simulated the fringe parallax.

The equivalence to time-averaged holography (Section 3.4) consists of one single activation of the ink with a duration of many vibration periods. The darkening properties of the ink should in this case be time dependent so that those parts on the object surface that have the longest exposure time to the ink become darkest.

5.5.3 The moiré fringe-forming process in empty space

There exists one alternative method of comprehending the moiré pattern analogy which might be more difficult to visualize but the reward is much higher for those who succeed.

Let us suppose that the imaginary ink between the ellipsoidal shells, when activated, marks not only the object surface but the three-dimensional space itself. Thus, the first recording is represented by one activation so that a local reference space fixed to the object is marked by the ellipsoids. Let this local reference space between the two exposures move with the object. The second exposure represents a second activation of the ink existing between the ellipsoidal shells of the holo-diagram. The points A and B are all the time fixed to the stationary points of illumination and observation, respectively.

Thus the local reference space is marked twice by the ellipsoids of the holo-diagram and the result is a three-dimensional moiré pattern. Interference fringes are seen where these dark moiré patterns intersect the object and the resulting pattern will be exactly the same as described by the first method of comprehending the moiré effect which was put forward. Our last method, however, has got one great advantage in that it visualizes the shape of the interference fringe surfaces in empty space and how the three-dimensional shape of the object influences the observed interference pattern.

5.5.4 Translation

The moiré pattern method described in the preceding sections is, of course, only a theoretical experiment that cannot be realized in practice, but it certainly can be used to simplify the understanding of hologram interferometry. However, if some limitations on the exactness of the analogy are accepted, the proposed method can be used in a very practical way.

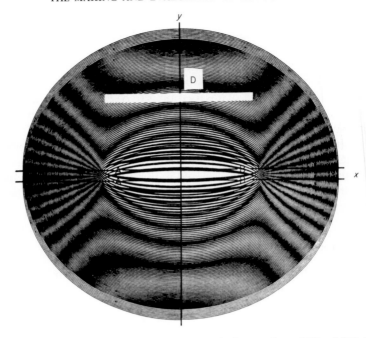

Fig. 5.15. Two transparent copies of a pattern similar to that of Fig. 1.24(a) have been placed on top of each other so that A and B coincide. Thereafter one transparency has been translated in the y-direction. Any object in a corresponding holographic set-up that is translated in the y-direction between the two exposures will get a fringe pattern represented by the intersection of the moiré pattern and a cross-section of the object.

Consider a difference in path length for light reflected from adjacent ellipsoidal shells of, say, 10 000 times longer than 0.5λ. For $\lambda = 0.6328\,\mu\text{m}$, this will give a difference in path length of about 3 mm. Examining only the intersections between the ink-filled ellipsoidal shells and a flat surface, i.e. a cross-section through the axes of the ellipsoids (as demonstrated in Fig. 1.24(a)), two transparent copies are made of a diagram similar to that of Fig. 1.24(a) and one is placed on top of the other so that A and B coincide.

With these two transparencies representing two sets of primary fringes, a simulation of hologram interferometry is easily performed. For example, move one transparency in the y-direction (Fig. 5.15). The resulting moiré pattern (the secondary fringes) is an almost exact copy of the holographic interference pattern that would be formed on a large surface that is displaced in the y-direction and placed just under and almost in contact with A and B. The sensitivity of this moiré method is only 10^{-4} times that of its holographic analogy, and therefore some slight error could be introduced in the vicinity of A and B. (To form the same number of fringes the displacement has to be 10^4 times larger.) The main structure of the fringe pattern is, however, easily studied; moreover it is easy, for example, to find out that a long object placed at D will have a surprisingly constant sensitivity to displacement in the y-direction over its whole length.

The two transparencies can just as easily simulate the displacement in any arbitrary direction in the (x, y)-plane. They can also simulate rotation around any point in that plane. To find the fringes formed at other cross-sections of the ellipsoids, other transparencies have to be made. It should be pointed out that the intersection of the set of ellipsoids and any surface forms a zone-plate that diffracts light from A to B. (This type of zone-plate is not restricted in having a flat surface but could have any arbitrary curvature.)

The secondary fringes of Fig. 5.15 represent the solutions to

$$K = \frac{2|y^2 + (Lx)^2|^{1/2} x|y^2 + (L-x)^2|^{1/2}}{y|y^2 + (L+x)^2|^{1/2} x|y^2 + (L-x)^2|^{1/2}}, \tag{5.33}$$

where K is an integer and the x-values of A and B are $+L$ and $-L$, respectively.

The pattern of Fig. 5.15 is made by a translation in the y-direction of one of the transparencies and corresponds to the diagram described in Chapter 4 (Fig. 4.26). The displacement of the translated set of primary fringes is constant over its whole area and therefore the secondary fringes at the same time represent the loci of constant sensitivity corresponding to Fig. 4.26. Pattern III(a) in Fig. 5.20 is made by a translation in the x-direction of one of the transparencies and this corresponds to the diagram of Fig. 4.22.

When the displacement of the object's surface in a hologram is a combination of rotation and translations in both the x- and y-directions, the mathematical solution is complex and rather complicated to work out. However, all those fringe patterns that are possible can be found using only the two transparencies. Those two sets of primary fringes thus represent hundreds of diagrams that can be used for the study of the main structure of the fringe patterns and the variations in sensitivity to object displacements in any direction over the whole area under study.

Sollid has described an experiment in which a flat surface is translated, first parallel to the normal of the surface and then in the plane of the surface. The displacement is equal in both cases, and the arrangement is such that the projection of the displacement on to the bisector of the directions of illumination and observation is also equal for the two experiments. Sollid points out that these two experiments are equivalent as far as the existing theory for reducing fringes to displacements is concerned. Yet in the first experiment there are only a few fringes to interpret; in the second experiment there are a great number of fringes.

Using the two transparencies the equivalent moiré experiment is easily

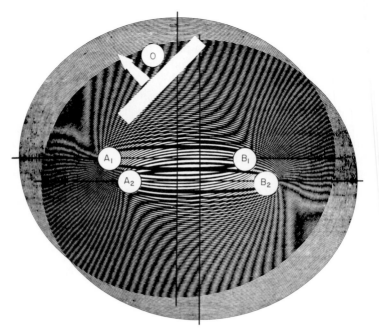

Fig. 5.16. Similar to Fig. 5.15 except that in this case the object has been translated parallel to the normal of its surface. Only two fringes are formed.

performed, and it produces exactly the result described. Only those fringes can be seen that are positioned on such parts of the surface of the object that can be illuminated from A and studied from B. The hologram interference fringes are represented by the intersection of that surface of the object and the secondary fringes. Figure 5.16 demonstrates a translation parallel to the normal to the surface (two fringes) and Fig. 5.17 a translation of the same amplitude parallel to the surface plane (12 fringes). In Figs 5.16 and 5.17 the direction of increasing sensitivity is visualized in a very concentrated way.

5.5.5 Rotation

Figure 5.8 is a photograph of the fringe pattern formed on the surface of a disc that was given a small rotation between the two exposures of the hologram. The disc is seen from an oblique angle. The centre of rotation is marked by a cross (+); the point at which one of the imaginary ellipsoids is tangential to the disc surface is marked by a small circle (o).

Figure 5.18 demonstrates the moiré equivalence to that experiment. The sets of primary ellipses represent the locus of intersection of the ellipsoids and the flat surface of the disc. The two transparencies have been rotated with

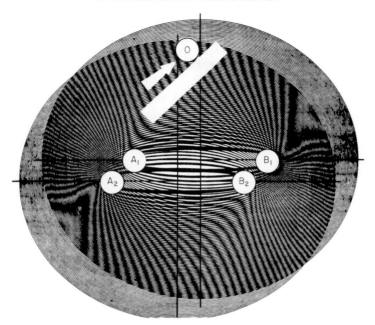

Fig. 5.17. Similar to Fig. 5.15 except that in this case the object has been translated in the plane of its surface. Twelve fringes are formed in spite of the fact that the translation and its projection on the bisector of the beams of illumination and observation are the same as those of Fig. 5.16.

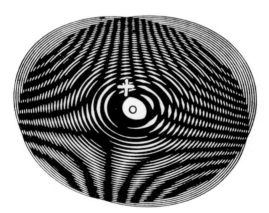

Fig. 5.18. The moiré pattern analogy to the interference pattern formed in the holographic experiment of Fig. 5.8. Two sets of identical transparencies, with the ellipses representing the intersections of the ellipsoids by the disc surface, have been placed one exactly on top of the other. Thereafter one set has been rotated around the cross (+) representing the disc's centre.

respect to one another around an axis marked by the cross (+). The tangent point is marked by a small circle (o). The secondary fringes of Figs 5.8 and 5.18 are surprisingly well correlated considering the great difference in sensitivity. One of the transparencies of Fig. 5.18 was rotated some 10 000 times more than the disc of Fig. 5.8.

Figures 5.14, 5.9 and 5.18 all agree with the photograph of the holographic fringe pattern seen in Fig. 5.8. Thus no less than three different methods have been proposed to explain the fringes formed by rotation, namely utilization of the intersections of the object by the ellipsoids, the analogy of reflections from motion trails, and finally the moiré pattern analogy. All these methods are derived from one basic idea regarding the ellipsoids of the holo-diagram. The reason why so many pages have been spent on the study of rotations is that every type of rigid in-plane motion can be represented by a rotation. Translation is just a special case in which the axis of rotation is positioned at an infinite distance.

5.5.6 Examples

To date the author has tested the moiré equivalence some 20 times and never found any marked discrepancy between the moiré pattern and the corresponding interferometry fringe pattern. Figure 5.19 shows those cross-sections of the ellipsoids that have been studied. The corresponding moiré patterns are seen in Fig. 5.20. Each pattern of Fig. 5.20 is marked in the right-hand corner by figures relating to the different cross-section of Fig. 5.19 that has

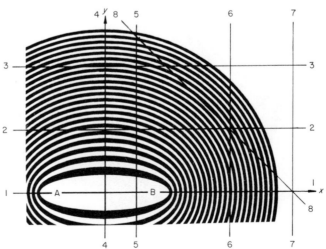

Fig. 5.19. The lines numbered from 1 to 8 represent the cross-sections through the ellipsoids that have been calculated and used as primary fringes in moiré pattern experiments.

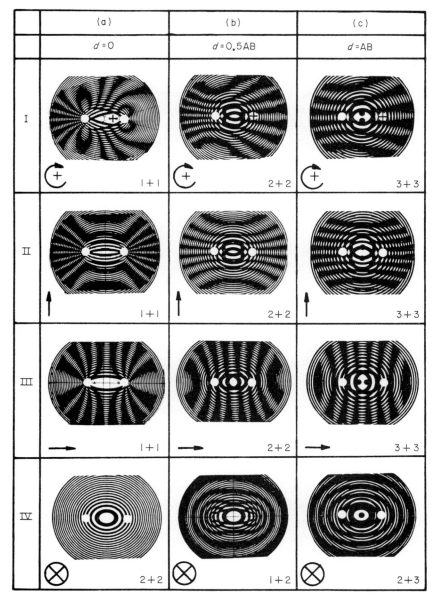

Fig. 5.20. The basic moiré patterns representing interference patterns formed on surfaces that have been given a small displacement between the two exposures of a hologram. The surfaces are supposed to be placed at a distance d behind the plane of the holo-diagram. A and B are fixed while the surfaces are displaced. I, rotation; II and III, translation in the plane of the paper; IV, translation normal to the paper. (a) Surface in the plane of the holo-diagram; (b) surface at a distance of 0.5AB behind the plane of the diagram; (c) surface at a distance of AB behind the plane of the diagram.

been used for producing the moiré fringes. An arrow indicates the direction of displacement.

All the patterns of Fig. 5.20 are similar to those predicted in Section 5.4. The points A and B are represented by white dots. I(a) demonstrates the pattern formed on a flat surface placed just behind A and B ($d = 0$) and rotated around the line through the cross and perpendicular to the surface. This pattern has been partly verified holographically. I(b) and I(c) demonstrate the patterns formed on a surface placed at distances 0.5AB and AB, respectively, behind A and B. In these two cases the centre of rotation was the normal to the surface through B. Both patterns have been verified holographically.

The patterns of rows II and III are the results of in-plane translation in a vertical and a horizontal direction respectively. Patterns II(a) and III(a) can only partly be studied holographically; patterns II(b), II(c), III(b) and III(c) are easy to verify.

Pattern IV(a) has no secondary fringes at all. This fact demonstrates that a plane passing through A and B is insensitive to a translation in a direction normal to its surface. The two primary patterns are identical as they represent two parallel cross-sections made at either side of the axis AB and at the same distance from this axis. This insensitivity has been verified holographically.

The two sets of circles around A and B that can just be visualized in IV(b) are rather difficult to see in the holographic image. A small fixed aperture has to be used for the observation because the fringes move very rapidly between A and B when the observer moves his head. The pattern of IV(c) is again easy to verify in a hologram.

The patterns of Fig. 5.20 represent the hologram fringe patterns formed on surfaces of the object placed in the vicinity of A and B. The further it is from A and B the more each cross-section of the ellipsoids resembles an ordinary circular zone-plate. Outside a sphere with radius AB and centre half-way between A and B, the primary pattern could well be approximated into Fresnel zone-plates. Thus in-plane translations anywhere outside this sphere will produce approximately straight fringes perpendicular to the direction of translation. Translations normal to the plane will produce approximately circular fringes, and if the centre is far from the line of sight these fringes also might give the impression of being straight lines.

There are two slight drawbacks to the moiré pattern method. One disadvantage is that the transparencies must be deformed to demonstrate the pattern resulting from deformation or extension of the object surface. The other disadvantage is that the principle as described does not give a simple demonstration of how the fringes move around and change as the observer moves his head (the dynamic properties of the fringes). Therefore it is difficult to demonstrate the possibility of discriminating between the different types of object motion that might produce identical stationary patterns.

A method is available to advance the uses of the moiré pattern analogy to include the motion of fringes as the observation angle is changed. This method, however, is rather complicated to use and will not be dealt with here. Instead the analogy of reflections from motion trails is recommended.

5.5.7 Analogy or identity

The method described happens to be an optical solution (moiré pattern) to an optical problem (fringe formation in hologram interferometry). It must, however, be pointed out that even though it gives an analogical result it does not in any way pretend to copy the nature of the formation of interference fringes. It is the belief of the author that a moiré pattern analogy could be used to solve other mathematical or physical problems where it is valuable to find the solution in the form of a "live" diagram.

By considering rotation and translations in two directions as variables that can be used as inputs in a moiré diagram, a "*live*" diagram is formed, and the curves of this diagram will twist and change in shape as the inputs are changed. In this way it is sometimes possible to find solutions to rather complex problems that would be difficult to visualize in other ways.

"Does there exist a moiré that represents the solution to any complex mathematical problem based on a number of simple relations? How should the primary fringes be calculated? These are some of the questions the answers to which might perhaps open up new possibilities of visual interpretation of complex mathematical calculations."

Those were the words accompanying the first publication of the moiré pattern analogy to hologram interferometry. Since then the author has changed his mind about the moiré pattern method being a mere analogy. As described in Section 5.5.1 there are good reasons to believe that the described moiré fringes are not only analogous but *identical* to the fringes formed in holographic interferometry.

The only difference is the scaling factor caused by the fact that it is impossible to draw a diagram with a set of ellipses with separations of the order of the wavelength of light. This scaling factor causes the diagram to be displaced much more than the object measured interferometrically. The disturbing influence of this factor on the pattern could be found by studying how much the main structure of the pattern changes as the displacement is doubled. No such influence has yet been seen when the moiré method is used in the way described in the text. Therefore it is concluded that the scaling factor has been of no important influence. This conclusion is, of course, further confirmed by the similarity between Figs 4.26 and 4.22, which are based on mathematical calculations, and patterns II(a) and III(a) in Fig. 5.20, which are based on the moiré pattern analogy.

One question remains, however. Are there really any fringes on the object representing the intersections by the ellipsoids? The author believes that that question should be answered in the affirmative. The dark fringes of a preliminary pattern are fringes of blindness, fringes representing zero information. A point localized in such a dark fringe will produce no difference at the point of observation whether it exists or not. Such areas of blindness exist in hologram interferometry.

When the reference beam and a signal beam from an object point intersect at the hologram plate image-carrying fringes are formed. If the object beam is weaker than the reference beam, destructive interference will appear at some points so that the resulting intensity is lower than that of the reference beam alone. At other places there will be constructive interference resulting in increased intensity (see Fig. 1.20).

For a special phase angle in between these extremes there will be no resulting change of intensity at all, whether the object point exists or not.

The object points that in this way are not recorded on the point of the plate under examination are all situated along the ellipsoids of the holo-diagram. These ellipsoids therefore represent the fringes of blindness at the object's surface which use the moiré effect to produce the interference fringes seen in hologram interferometry.

5.6 Speckle methods

It has been demonstrated mathematically that the fringes of holographic interferometry can be derived solely from the speckle motion at the plate. This was verified for tilt (equation (4.23)) and also for in-plane translations (equation (5.22)). In both cases the twin speckles at the plate produce Young's fringes that are projected on to the image during reconstruction. There is, however, one strange conclusion to be drawn from this method of explaining the interference fringes. As the interference effect of the reference beam is not included in the evaluation it has no influence on the measuring process. The most extreme conclusion which can be drawn from this fact could be proposed in the following bold statement: *the phase determination effect of the reference beam is not needed in holographic interferometry*. The single use of the reference beam is that it produces the holographic image on to which the fringes are projected.

When the object under examination is differently displaced and deformed at different parts, then the holographic imaging process is needed to sort out the different Young's fringe patterns and to project them at their proper place in the image. Looking at the measuring process in this way it can be understood that it ought to be possible to exchange the holographic method for other imaging techniques.

5.6.1 Speckle photography

There already exists one method of measurement which proves the statement in the preceding section: speckle photography. The object is focused on to a photographic film or plate using an ordinary high quality camera.[9] If the object makes an in-plane motion (d_i) between the two exposures, two images will be formed on the film (Fig. 5.21(a)). The object should be illuminated by laser light and thus every speckle on the object surface will be doubly exposed because they move with the object. If the demagnification is n, the separation of the twin speckles on the film will be nd_i. When the processed film is illuminated by a thin laser beam Young's fringes will be formed which could be projected on a screen at a distance L_2 (see Fig. 5.21(b)). The separation (e) of the Young's fringes will be given by

$$e = \frac{\lambda}{2 \sin \beta} \sim \frac{\lambda L_2}{n d_i}. \tag{5.34}$$

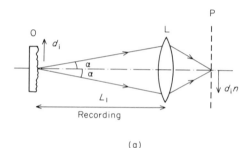

(a)

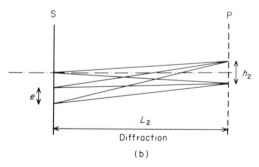

(b)

Fig. 5.21. (a) Speckle photography. The object (O) at the distance L_1 is focused on to a photographic plate (P) by the camera lens (L). As the object is given an in-plane displacement (d_i) the speckles at its image will move the distance $h = nd_i$, where n is a demagnifying factor.

(b) When the twin speckles on the photographic plate (P) are illuminated by laser light, Young's fringes with a separation of e are projected on to a screen (S) at the distance L_2.

The smallest value of d_i that could be resolved is equal to the resolution (r) of the camera (equations (1.13) and (3.1)):

$$r = \frac{1.22}{2\sin\alpha} \sim \frac{1.22\lambda L_1}{D},$$

where D is the diameter of the camera lens while L_1 is its distance from the object. Thus the same resolution has been found when the camera was used directly (equation (1.19)) or when the speckle method was used (also equation (1.19)) or when a dynamic observation was carried out by moving the eye behind the hologram plate (equation (4.11)). If the method of studying one point on the object at a time by looking at the separation of the Young's fringes is not satisfactory, a more sophisticated type of optical filtering can be used (Section 5.2.4).

In this way it is possible to get the image of the total object covered by fringes looking similar to holographic fringes, but with the great advantage that it is also possible to change the magnitude and direction of the sensitivity by changing the position of the filtering pinhole. Another great advantage is that the method has no sensitivity to out-of-plane motion, where such motion is defined as movement along the line of sight. The reason is that when the surface of the object is properly focused by the lens, the speckle rays caused by tilt will all be focused at the same point on the film. This fact is based on the definition of the focusing effect. All the rays from one point on the surface of the object are directed by the lens to one single point at the image plane, independently of their angle of illumination.

5.6.2 Mathematical unification of speckle photography and hologram interferometry

If we want to measure out-of-plane motion (e.g. tilt) instead of in-plane motion, then the object must be defocused by the speckle camera (Fig. 5.22). Assume that the object is tilted so that one speckle is displaced by a distance h_1 (equation (4.13)) on the lens. The speckle motion (h_2) on the plate will be given by

$$h_2 = h_1 \Delta b/b, \tag{5.35}$$

where b is the distance between lens and image plane while Δb is the distance between the image and the film. Thus $\Delta b/b$ is a measure of the defocusing.

From Fig. 5.22 it can be seen that if $\Delta b = 0$ there will be no sensitivity to out-of-plane motion (tilt). If, on the other hand, $\Delta b = b$ the speckle motion is identical to that of holographic interferometry. Combining equation (4.21) with equation (5.35) the following result is obtained for the Young's fringes formed on a screen when a laser beam is diffracted by the processed

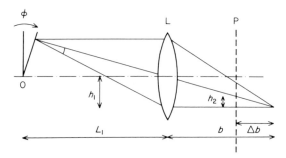

Fig. 5.22. Unification of speckle photography and holographic interferometry. O is an object tilted at the angle ϕ between the two exposures. h_1 is the speckle motion on the lens (L) while h_2 is the speckle motion on the photographic plate (P) which is separated by Δb from the focused image at the distance b. Projecting a laser beam through the plate (P) with its speckle separation of h_2 produces Young's fringes on the screen S at the distance L_2 (see Fig. 5.21(b)). The separation e of these Young's fringes will be equal to that of holographic interferometry if L_1 is equal to L_1 and Δb equal to b.

photographic film (Fig. 5.21(b)):

$$e = \frac{\lambda \cos \beta}{\delta \alpha (\cos \alpha + \cos \beta)} \frac{L_2}{L_1} \frac{b}{\Delta b}, \tag{5.36}$$

where e is the separation of the projected Young's fringes, α is the angle separating the normal to the surface of the object and the direction of illumination, $\delta \alpha$ is the object tilt, β is the angle separating the normal to the surface of the object and the direction of observation, L_1 is the distance from the object to the lens, L_2 is the distance from the film to the screen, b is the distance from the lens to the image plane, and Δb is the distance from the image plane to the film.

If $L_2 = L_1$ and if $\Delta b = b$, equation (5.36) represents the speckle motion on an ordinary hologram plate. It has already been proved that the holographic interference fringes can be evaluated correctly from this speckle motion. Thus equation (5.36) provides a mathematical link between speckle photography and holographic interferometry.

5.6.3 Sandwich holography

Just as the tilt of a sandwich hologram can be used to compensate for the speckle motion caused by tilt it can also compensate for the speckle motion caused by an in-plane translation of the object. The speckle motion h caused by in-plane motion d_i has already been evaluated in equation (5.20):

$$h = d_i \left(\frac{L \cos^2 \alpha + R \cos^2 \beta}{R \cos \beta} \right).$$

To compensate for this separation of the speckle twins the sandwich hologram has to be tilted at an angle ϕ_2 (equation (3.10)). By combining equations (5.3) and (3.10) the following expression can be obtained:

$$d_i = \phi_2 \frac{dR \cos \beta}{n(L \cos^2 \alpha + R \cos^2 \beta)}, \tag{5.37}$$

where d_i is the required in-plane displacement, ϕ_2 is the sandwich tilt needed for fringe compensation, d is the emulsion separation, R is the distance to the source of illumination, L is the distance to the point of observation, α is the angle between the direction of illumination and the normal to the surface of the object, β is the angle between the direction of observation and the normal to the surface of the object, and n is the refractive index. This equation can be approximated to such small angles ϕ_2 that $\sin \phi_2 \sim \phi_2$ and that the motion of the reconstructed image (equation (4.24)) is negligible.

Finally, let us compare sandwich holography to conventional holography for the separation of out-of-plane motions (e.g. tilt) from in-plane motions. In ordinary double-exposure holography the in-plane component can be found by moving the eye behind the hologram plate and observing the motion of the fringes. From this observation it is possible to evaluate the in-plane motion and calculate the number of fringes it would have caused. By subtracting these fringes from the original fringe system it is finally possible to find the fringes caused by tilt alone.

Using sandwich holography the evaluation is less complicated. The hologram is tilted until the fringes appear fixed to the object as the eye is moved behind the plate. During this tilt the fringes caused by in-plane motion are automatically eliminated so that the remaining stationary fringes represent the tilt alone.

The resolution of the method is, however, limited by the fact that the hologram tilt resulting in stationary fringes is not so well defined as the corresponding tilt resulting in no fringes at all.

References

1. E. Aleksandrov and A. Boch-Bruevich. Investigation of surface strain by the hologram technique. *Sov. Phys. tech. Phys.* **12**, 258 (1967).
2. K. Stetson. A rigorous theory of the fringes of hologram interferometry. *Optik* **29**, 386 (1969).
3. T. Matsumoto, K. Iwata and R. Nagata. Distortionless recording in double-exposure holographic interferogram. *Appl. Opt.* **12**, 1660 (1973).
4. J. Butters. Applications of holography to instrument diaphragm deformations and associated topics. In *Engineering Uses of Holography* (E. R. Robertson and J. M. Harvey, eds). Cambridge University Press, Cambridge (1970), p. 151.

5. J. Vienot, C. Froehly, J. Monnet and J. Pasteur. Hologram interferometry surface displacement fringe analysis as an approach to the study of mechanical strains and other applications to the determination of anisotropy in transparent objects. In *Engineering Uses of Holography* (E. R. Robertson and J. M. Harvey, eds). Cambridge University Press, Cambridge (1970), p. 153.
6. N. Abramson. The holo-diagram. III. A practical device for predicting fringe patterns in hologram interferometry. *Appl. Opt.* **9**, 2311 (1970).
7. N. Abramson. Moiré patterns and hologram interferometry. *Nature, Lond.* **231**, 66 (1970).
8. N. Abramson. The holo-diagram. IV. A practical device for simulating fringe patterns in hologram interferometry. *Appl. Opt.* **10**, 2155 (1971).
9. E. Archbold and A. Ennos. Displacement measurement from double-exposure laser photographs. *Opt. Acta* **19**, 253 (1972).

6

Fringe reading and manipulation

The difficulties of fringe reading are emphasized in the demonstration given in Fig. 6.1. The object, consisting of a vertical steel bar fixed at its lower end by two screws to a stable mount is shown on the left (Fig. 6.1(a), (b)). The top of the bar is supported by a point contact which slides and cannot carry any bending moment.

Between the two exposures of a hologram a force (F) is applied at the middle of the bar. The result is a bending of the bar as seen, albeit exaggerated, in Fig. 6.1(a) so that it intersects a number of imaginary interference surfaces (Section 3.2.2 and Fig. 3.7). To simplify the evaluation it is supposed that illumination and observation are both carried out in a direction that is normal to the object's surface and parallel to the force.

From Fig. 6.1(a) it is easy to draw Fig. 6.1(b), which represents the object seen from the front and covered by holographic interference fringes. All that is necessary is to draw one horizontal line from each intersection of the object by the flat interference surfaces. Thus there is no doubt that the deformation of Fig. 6.1(a) will result in the fringe pattern of Fig. 6.1(b).

When holographic interferometry is used for measurement, however, the problem to be solved is the reverse one. Given the fringe pattern of Fig. 6.1(b) we have to find the object deformation illustrated in Fig. 6.1(a). This task is very difficult indeed as there exist, theoretically, an infinite number of solutions. If the problem is limited to the incorporation of only those situations in which the object deformation is not only continuous but also as simple as possible, then four elementary cases can be identified.

The two elementary solutions are given in Fig. 6.1(c), where I–II–III represents the original deformation while I–VI–III represents its mirror image. This misinterpretation of the deformation is caused by the holographic method's lack of information about the sign of displacement. This inability to distinguish between forward and backward depends in its turn on the fact

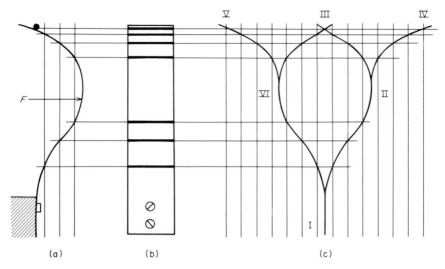

Fig. 6.1. The rose of error. There exists only one single fringe pattern (b) corresponding to the deformation mode (a), but there exist at least four deformation modes that would all produce the fringe pattern shown in (b). If one of the false modes is chosen, the bending moment might be misinterpreted as being zero (I–II–IV) instead of being a maximum (I–II–III), while a mistake of the direction of deformation (e.g. VI instead of II) produces an error of 200%.

that an ordinary holographic plate does not remember which exposure came first.

Then there exist two more simple possible solutions to the pattern of Fig. 6.1(b). The object could have been deformed along I–II–IV or along I–VI–V. These two misinterpretations are also caused by the lack of information regarding direction. The result is that the area of large fringe separation is erroneously interpreted as a point of inflection (no bending) instead of a point of high curvature (large bending).

The errors implied in Fig. 6.1(c) are by no means insignificant. The deflection error between the points II and VI is no less than 200%. The deflection errors at IV and V are still larger, the true displacement is zero while the measured displacement is double the value of points II and VI. Finally, the true bending moment at the point of application of the force F is close to a maximum, whereas the corresponding bending moment in the case of I–II–IV is zero at the point of inflection.

From these examples it can be understood that it is of no use to include exact values of the directions of illumination and observation in complicated and precise equations if there is already an error of 200% in the reading of the fringes!

Thus it is of great importance to find ways to elicit the correct information from the four possible interpretations demonstrated by the "rose of error" of Fig. 6.1(c). There are two possibilities for doing this: either known movement of the object can be introduced between the two deformation situations to be studied or such a movement can be introduced in an artificial way during reconstruction. The latter method could, for example, be accomplished by using real-time holography (Section 3.2) or any other method producing two images that are recorded separately (Sections 3.3.2 and 3.5).

The following sections will concentrate on the advantages of fringe manipulation during reconstruction. Special emphasis will be given to the use of sandwich holography (Section 3.5) simply because that is the method which the present author has found to be most useful in practical work.

6.1 Compensation for unwanted rigid body motion

During most practical work with holographic interferometry the whole holographic set-up is slightly deformed by the force used in the loading of the object. The required fringe pattern can therefore be drowned in the pattern caused by the rigid motion of the object. In other cases the intention could be that of studying the local deformation of a small part of a large object which is not only subject to rigid body motion but also to deformation. In that case it again can become very difficult to distinguish the fringes caused by the required deformation from those that deal with the often much larger unwanted motion.

By the use of a television camera connected to a suitably programmed computer it should be possible to distinguish the required fringe pattern from the unwanted one and to eliminate the latter. Such a system, however, would become extremely complicated, especially as the uncertainty about the sign of the unknown motion has to be overcome. Thus it has been found, instead, to be advantageous to carry out the compensation process as early as possible during the reconstruction stage.

6.1.1 Interferometric compensation

The most direct method which could be used to compensate for motion of the object would be to move the object itself during its observation by the use of real-time holography (Section 3.2): at the same time the force is applied to the object to produce the required deformation, direct compensation for unwanted object motion could be instigated. This could be done by the use of micro-positioning devices, e.g. utilizing piezoelectric materials.

Such a method would be costly and complicated but it would undoubtedly produce almost ideal compensation. Its main disadvantage would probably be that the real-time set-up has to be left unchanged during the whole evaluation stage. If the object was too large and too heavy to be moved by micro-positioning devices, an optical component could be moved instead to produce the compensation. Sometimes, and very simply, the motion of a parallel piece of glass in the reconstruction beam can be utilized.

In order to carry out the compensation process *after* the two holographic recordings have been made, the double reference beam method of Section 3.3.2 can be used. By moving one reconstruction beam in relation to the other rigid body motions of the object can be compensated for. After testing various different methods, however, the author found that the sandwich holography method (Section 3.5) was the easiest to use for compensation for unwanted movements of the object of 1 mm and higher.

It has already been verified that if an object is tilted (equation (4.20)) or translated (equation (5.20)), the speckles on the hologram plate are displaced. When there is a double exposure twin speckles are recorded and the Young's fringes caused by these twin speckles are identical to the holographic interference fringes (Sections 4.2.8 and 5.4.1). If the distance (h) separating the twin speckles can be monitored during reconstruction, it is evident that the holographic interference fringes can be changed.

When the two plates of sandwich holography are so manipulated during the reconstruction that the transverse separation of the twin speckles becomes zero, then the number of holographic interference fringes will also be zero. As the hologram plates are rigid it is possible to set only one single, constant, twin speckle separation at zero and thus to compensate for only one particular constant displacement over the whole object. Therefore sandwich holography is especially useful in compensating for rigid body motion.

6.1.2 Compensation by the moiré effect

There exists an alternative explanation of the fringe manipulation process which gives the same result as already discussed but which also produces one additional solution. Every point on the object produces, together with the reference beam, a set of Young's fringes on the hologram plate. Thus the image-carrying fringes (primary fringes) on the plate represent its intersection by a set of rotationally symmetric hyperboloids (Section 1.4.1, Fig. 1.24(a)).

If one point on the plate is examined, e.g. a point on a bright primary fringe, no dark fringe will pass over this point as long as the object point under scrutiny is moved along one particular ellipsoid of the holo-diagram (Section 1.4.1, Fig. 1.24(b)). When, for example, the object point passes through three ellipsoids between the two exposures then three dark hyperboloids will pass through the point on the plate. Thus, for every ellipsoid in the image plane there exists one hyperboloid in the plane of the hologram plate (for experimental verification see Section 7.5.2).

It has already been proved (Section 5.5) that the holographic fringes on the image of the object can be represented by the moiré effect of the ellipsoids. It is also evident that a double exposure produces a moiré effect of the hyperboloids at the hologram plate. As there is a one-to-one relationship between the ellipsoids and the hyperboloids a manipulation of the moiré pattern at the plates influences the moiré pattern on the object. If the displacement of the two sets of hyperbolas at the hologram plate is eliminated, then the holographic fringes on the object also disappear.

This method of explaining the function of the sandwich hologram method has the advantage that it clearly demonstrates that there should exist two procedures for the manipulation of the fringe pattern seen on the object: either by a moiré effect at the hologram plate or by a moiré effect at the image of the object.

In the following examples both types of manipulation will be given and it will be indicated that the moiré effect in the image plane has some grave limitations: it cannot regain the information lost because the fringes were too closely spaced to be resolved, neither can it regain lost information concerning the direction of the displacement.

6.1.3 Experiments

The object consisted of three vertical steel bars,[1] each one fixed to a rigid steel frame by two screws at its lower end. The top ends of the left-hand and the right-hand bars were free but the middle bar was also supported at its top by a point contact which slides and cannot carry any bending moment—as seen in Fig. 6.1(a).

Between the two exposures of the hologram a force was applied at the middle of the bars where a small dark spot can be observed in Figs 6.2–6.4. The force was applied perpendicularly to the surface of the object but its direction was different for the different bars. The illumination and the observation were both made from a direction normal to the surface and from distances that were more than five times the dimensions of the object. The reference beam was diverged from a distance of more than ten times the object's dimensions. The resulting reconstruction is seen in Fig. 6.2.

In Fig. 6.2 rigid body motion of the entire object, frame and all, has disguised the required information concerning the deflections of the bars in relation to the supposedly fixed frame. Apparently the applied force not only produced the requisite deformations of the bars but it also tilted the whole object, so there is no fixed reference frame. Thus the fringe patterns caused by the deflection of the bars have been added coherently to the fringe pattern caused by rigid body motion; the relevant information is therefore severely disguised by spurious fringes.

If the reconstruction of Fig. 6.2 had been made from an ordinary hologram

Fig. 6.2. Rigid body motion of the entire object has disguised the information concerning deformations. If this hologram had been an ordinary double exposure it would have been judged a failure.

it would have been considered a complete failure. However, using the concept of Young's fringes of observation, the conclusion can be drawn that the speckles at the hologram plate caused by rigid body motion of the whole object have moved along a line that is perpendicular to the fringe pattern at the object frame. Any method by which it is possible to compensate for this displacement of the recorded twin speckles can be used to eliminate the unwanted fringes.

As the image of Fig. 6.2 was reconstructed from a sandwich hologram it is possible to manipulate the separation of the recorded twin speckles either by shearing one plate in relation to the other or by tilting the bonded sandwich hologram (Fig. 3.24). The latter method was chosen and, while the reconstructed object was examined, the sandwich hologram was tilted until the frame became fringe-free.

The result is seen in Fig. 6.3, from which the deflections of the three bars can be observed in relation to the frame just as well as if the total object, frame and all, had remained stationary between the two exposures. To find the deflection of the top of the left-hand bar it is a simple matter to count how many dark fringes have to be crossed as the eye travels between its fixed lower end and its top (ten fringes). The same evaluation can be carried out on the right-hand bar (eight fringes). But what has happened to the middle bar?

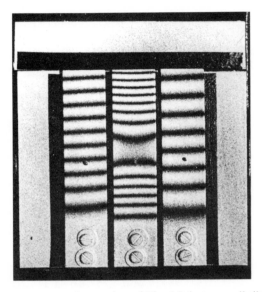

Fig. 6.3. The same reconstruction as that of Fig. 6.2, but a small tilt of the sandwich hologram has eliminated the fringes on the rigid frame. The fringe pattern now correlates well with the expected deformation. The object consists of three bars, the lower ends of which are screwed to a rigid frame. The middle bar is also supported at the top end. Between the two exposures forces were applied at the middle of the bars. The right-hand bar is deflected away from the observer. The other two bars are deflected towards the observer.

6.2 Evaluation

Because the information about the direction of deflection is missing in the reconstruction of Fig. 6.3, it by no means presents the whole picture regarding deformations of the object. Most obviously, it does not show whether the bars are deflected forwards or backwards. It is therefore impossible to tell whether the relative displacement of the left-hand and the right-hand bars is the difference (two) or the sum (18) of their individual number of fringes.

The fringes at the middle bar look strange about half-way up. Is that caused by a point of maximal deflection or by an inflection point? Should the fringes be counted as six plus eight or as six minus eight? To find the answers to these questions it is essential to be able to manipulate the fringes.

6.2.1 Direction of motion

By tilting the sandwich hologram so that its top is moved towards the observer the twin speckles from the tilted top half of the left-hand bar were

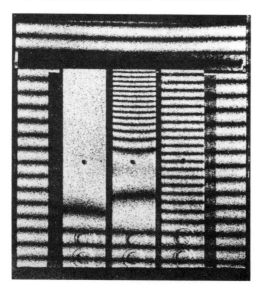

Fig. 6.4. The same reconstruction as that of Fig. 6.3, but the sandwich hologram has been tilted so that its top is moved towards the observer (Fig. 3.22). At a certain tilt angle (ϕ_2) the top of the left-hand bar (which was deflected by the angle ϕ_1 towards the hologram) is fringe-free.

aligned along the line of sight. Thus the spacing of the Young's fringes from that part of the hologram became infinite and it appeared fringe-free. From this result it is concluded that the left-hand bar was deflected in such a way that its top had moved towards the observer. (The first exposed plate was used as a rear plate, as described in Section 3.5.) The angle of tilt can be calculated from the sandwich angle using equation (3.10).

In Fig. 6.4 the right-hand bar is covered by more fringes than in Fig. 6.3, which proves that it was deflected in the opposite direction compared to the left-hand bar. The relative motion between the tops of the two bars is found by counting the difference in the number of fringes on the left-hand bar (four) and the right-hand bar (22). The result is a relative motion of 18 fringes, which is identical to one of the two possible results derived from Fig. 6.3. The explanation of the fringe behaviour exhibited in this experiment is seen in Fig. 6.5, where Fig. 6.5(a) represents the fringe formation of Fig. 6.3 while Fig. 6.5(b) represents that of Fig. 6.4. The result of sandwich tilt is represented in Fig. 6.5 by a tilt of the imaginary reference surfaces. It should be kept in mind that the angle ϕ_1 (see also Fig. 3.22) is infinitesimal.

Tilting the top of the sandwich hologram away from the observer results in the reconstruction shown in Fig. 6.6, which confirms the results of Figs 6.4 and 6.5.

6.2.2 Bending

Let us now study the middle bar. There is a point of maximum fringe separation at its centre, as shown in Fig. 6.3, while that point has moved downwards in Fig. 6.4 and upwards in Fig. 6.6. From these results the conclusion can be drawn that the assumed bending form indicated in Fig. 6.5 is correct. Thus it has proved possible to solve the problem of Fig. 6.1: of the four possible bending forms the single correct one has been selected. However, it is still rather difficult to visualize the actual bending curve of the middle bar. Figure 6.5 does not represent a bad solution, but it takes time to make and would be complicated to use for a large area covered by fringes caused by a complex deformation pattern.

If the bending radius is to be found, it is possible to measure the sandwich tilt needed to produce a tangent point between the bending curve of the object and the imaginary interference surfaces. This point can be observed in the reconstruction as the point of largest fringe spacing. By moving it up and down along the object (Figs 6.3, 6.4 and 6.5) the variation in the angle of tilt of the object can be calculated and from that the bending radius.

Another procedure is to measure the variation in fringe separation along the object and from that calculate the variation in tilt, which is a measure of

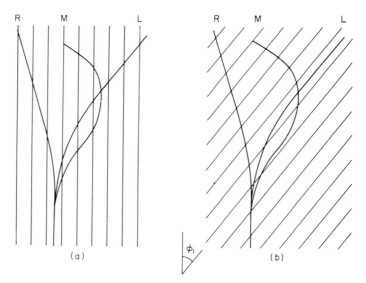

Fig. 6.5. The deformed bars of Fig. 6.2 are seen from the side as they intersect the interference surfaces of Fig. 3.7. R, M and L refer to the right-hand, the middle and the left-hand bars. In (a) there is a tangent point half-way up the middle bar. In (b) the interference surfaces have been tilted so that they become parallel to the left-hand bar. The tangent point of the middle bar has moved downwards. This situation represents the photograph in Fig. 6.4.

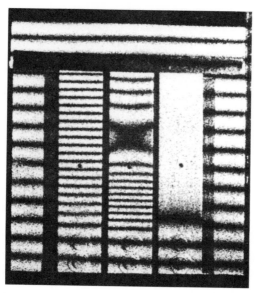

Fig. 6.6. Tilting the sandwich hologram in the opposite direction compared to that of Fig. 6.4 results in the right-hand bar becoming fringe-free while the fringe number of the left-hand bar is increased compared to that of Fig. 6.3.

the bending radius. It is, however, difficult to measure the position of a fringe with high accuracy and the inevitable error will be greatly magnified as the positions are used to find the fringe spacings and from these to calculate their difference, the derivative of the spacings. To solve this problem it is possible to use either the complex method of holographic heterodyning with two reference beams (Section 3.3.2) or the fringe manipulation procedure described below.

Let us go back to Fig. 6.3 and re-examine the middle bar: from earlier experiments it is known this bar was bent towards the hologram plate (Fig. 6.7(a)). The fringe formation process is visualized in Fig. 6.7(b). Had the bar been rotated around a vertical axis instead of being bent forward, the reconstruction would have shown its surface covered by vertical fringes (Fig. 6.7(c)). If *both* the motions of Fig. 6.7(a) and (c) had happened between the two exposures, the pattern shown in Fig. 6.7(d) would have been the result. The curve of Fig. 6.7(d) represents the intersection of the bent and rotated object by a single imaginary surface. Points on this intersection are found by marking the cross-over points of the vertical and the horizontal lines of the same number. The reason for this is that the object is now inclined to the interference surfaces, as seen in the projections at the top and the bottom of Fig. 6.7(c).

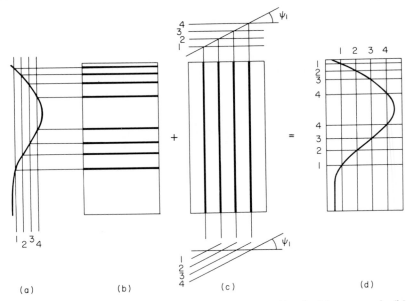

Fig. 6.7. The deformation of the object (a) is represented by the fringes seen in (b). In (c) it is demonstrated that a tilt of either the object or of the interference surfaces produces vertical fringes. The coherent combination of the fringes of (b) or (c) results in fringes that directly visualize the deformation mode of the object (d).

The bending curve constructed in Fig. 6.7(d) represents the single correct solution to the example discussed in Fig. 6.1; thus the introduced object rotation results in yet another method for the elimination of the erroneous bending forms. There is, however, one difficulty with the proposed method. It is rather difficult to introduce the small rotation of the object in between the two exposures, especially as that has to be done with such an interferometric precision that it produces no other unwanted motions. If, however, fringe manipulation after recording is used instead, then the method becomes practical and easy to handle.

A rotation of the sandwich hologram around a vertical axis produces a horizontal separation of the twin speckles and therefore has exactly the same result on the holographic interference fringes as a rotation of the object itself. Thus Fig. 6.7(b) represents the fringes caused by object deflection, while Fig. 6.7(c) represents the fringe pattern caused by either object rotation (projection at top) or by sandwich hologram rotation (projection below) around a vertical axis. Finally, Fig. 6.7(d) represents the fringe pattern resulting from the coherent addition of the patterns caused by either of these motions to that caused by the deflection.

Figure 6.8 shows the practical result of rotating the sandwich hologram of

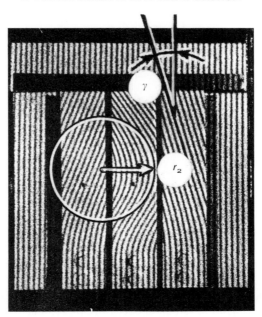

Fig. 6.8. The same image as that of Fig. 6.3 but the sandwich has been rotated by the angle ψ_2 around a vertical axis. The angle (γ) between the fringes and a vertical line is a measure of the angle of tilt of the object (ϕ_1). The radius of curvature of the fringes (r_2) is a measure of the bending radius (r_1) of the object, from which the bending moment and maximal strain of the object can be calculated. Observe the large information content of Fig. 6.8 compared to that of Fig. 6.3.

Fig. 6.3 around a vertical axis. The frame, to which it is intended to relate all the deflections, is covered by vertical fringes. The lower parts of the bars are also covered by vertical fringes which represent zero deflection. The larger the angle of tilt of the object, the larger is the angle of tilt of the fringes. Even the direction of the angle of tilt is inherent in the fringes. When the object is bent forward, the fringes are inclined to the right (left-hand bar); when the object is bent away from the observer the fringes are inclined to the left (right-hand bar). These relationships obviously depend on the direction of rotation of the sandwich hologram being such that it can be found from Fig. 6.7.

The rule is that a fringe inclination to the right represents a tilt of the object towards the observer if the rear plate was exposed first and if the sandwich hologram is rotated clockwise as seen when looking vertically downwards.

Another way to find the deflection of the right-hand bar of Fig. 6.8 is to follow one inclined fringe from the lower end to the top and to study how far to the left from a vertical line the fringe is deflected. If this horizontal deflection is expressed in terms of the number (n) of fringe spacings (s) the result will be eight, which is identical to the number of fringes seen in Fig. 6.3. From this result it is possible to find the relation between fringe tilt (γ) and object tilt (ϕ_1) when the object length is (l):

$$\phi_1 = n\lambda/2l, \tag{6.1}$$

$$\tan \gamma = ns/l. \tag{6.2}$$

Thus

$$\phi_1 = (\lambda/2s)\tan \gamma. \tag{6.3}$$

The curvature of the fringes is a measure of the change in curvature of the object caused by the loading between the two exposures. Thus the bending radius—and hence the bending moment—can be calculated from Fig. 6.8 without any need of finding the derivative of the fringe spacing. Let us, as an example, study the bending radius (r_2) for the middle bar.

The radius r_2 in Fig. 6.8 is about 11 horizontal fringe spacings (s) over a length of the bar that is equal to $2r_2$ or about half the length of the bar. It is obvious that in reality the bending radius is much larger than the fringe radius of Fig. 6.8. A schematic view of the situation in Fig. 6.8 is shown in Fig. 6.9(a), from which the radius of curvature (r_2) can be expressed in terms of the number of fringes (n) and is equal to 11. In reality the deflection (h) of that bar would be only $n\lambda/2$. Thus the true bending radius (r_1) is much larger. Using the Pythagorean theorem on the situation shown in Fig. 6.9(b), the following result is obtained:

$$r_1 = (r_1 - h)^2 + r_2^2. \tag{6.4}$$

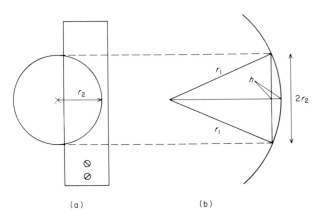

(a)   (b)

Fig. 6.9. (a) The fringe radius (r_2) of curvature of Fig. 6.8 is expressed in terms of the number of fringes. (b) Multiplying this number by 0.5λ gives the segment height (h) of the much larger radius (r_1) of deformation.

Fig. 6.10. The sandwich hologram is tilted around a horizontal axis from the position of Fig. 6.8 until the fringes become vertical at the right-hand bars. This situation of vertical fringes can be read with high accuracy, whereas it might be difficult to decide when an object is fringe-free, as in Fig. 6.6.

Now, if the small factor h^2 is neglected, the result is simplified:

$$r_1 = r_2^2/2h = r_2^2/n\lambda. \tag{6.5}$$

This result is true only if the photographic image of Fig. 6.10 is of the same size as the real object and if r_1 is much larger than r_2. From the bending radius the stresses and strains caused by the bending moment can be calculated if the modulus of elasticity and the inertia of the object's cross-section are known.

If r_2 is set equal to sn, where s is the horizontal fringe spacing which can be referred back to the tilt of a sandwich hologram, then the following equations are formed:

$$\phi_1 = \frac{d}{2Ln}\phi_2, \tag{6.6}$$

$$\phi_1 = \frac{d}{2Ln}\psi_2\tan\gamma, \tag{6.7}$$

$$r_1 = \frac{2Ln}{d}\frac{1}{\psi_2}r_2, \tag{6.8}$$

$$\sigma = \frac{d}{2Ln}\frac{\psi_2 Et}{2}\frac{1}{r_2}.$$ (6.9)

The symbols are defined in Figs 3.24 and 6.8, where ϕ_1 is the angle of tilt of the object, ϕ_2 is the angle of tilt of the sandwich, d is the thickness of the glass plate of the hologram, L is the distance from the object to the hologram plate, n is the refractive index of the hologram plate, ψ_2 is the angle of rotation of the sandwich around the vertical axis, γ is the angle of inclination of the fringe to the vertical axis, r_1 is the bending radius of the object, r_2 is the radius of curvature of the fringe, σ is the stress on the object, E is the modulus of elasticity of the object, and t is the thickness of the object.

These equations are limited to the study of displacements caused by a tilt of the object that is parallel to the direction of illumination and observation. It has also been assumed that the angle of tilt of the sandwich (ϕ_2) is so small that $\sin \phi_2 = \phi_2$ and that the motion of the reconstructed image (Section 4.2.8) can be neglected. The wavelength of light is not included in the equations because a compensation method has been used in which the interference fringes are only used to indicate non-zero conditions. For a more thorough theoretical study see Schumann and Dubas.[2]

Wherever possible the author recommends reading of the fringes as being preferable to measuring the angle of tilt of the sandwich because the former method is more direct and does not include the chain of possible errors caused by incomplete knowledge of the following factors: angle of tilt of the sandwich, plate thickness and variations thereof, refractive index of plate, and distance from the object to the hologram plate. Sometimes, however, it is necessary to use the angle of tilt of the sandwich for the evaluation, e.g. when the movement of the object is so large, and therefore the fringes so closely spaced, that they cannot initially be recorded.

It is often difficult to decide when an object is fringe-free because a variation in brightness caused by, say, a quarter of a fringe over a large area is very difficult to discern, especially if the brightness of the fringe-free object also varies over its surface. In such a case vertical fringes can be introduced at the reference frame of the object, as seen in Fig. 6.8. Thereafter the sandwich is tilted around a horizontal axis until the part under scrutiny is covered by vertical fringes, as is the right-hand bar of Fig. 6.10. In this way the fringe-free condition of Fig. 6.4 has been transformed to the condition of vertical fringes, which can be resolved with considerably higher accuracy.

6.2.3 Moiré pattern methods

It has already been pointed out that holographic interference fringes can be explained as a moiré effect either between two sets of hyperbolas at the hologram plate or between two sets of ellipsoids at the image plane. The former fringe pattern can be manipulated by moving one plate in relation to the other, as explained in Section 6.2.2.

It is also possible to manipulate the fringes in the reconstructed doubly exposed image by subtracting the fringe pattern which should have been formed if the object had undergone a particular change, e.g. rigid body motion.[3] Let us re-examine Fig. 6.8. This pattern corresponds to an object that is both deformed and rotated around a vertical axis. If the fringe pattern caused by rotation alone (the grid pattern formed on the undeformed frame) is superimposed on this image, the moiré effect will produce the pattern representing deformation alone, as demonstrated in Fig. 6.11 (cf. Fig. 6.3).

By tilting the added fringe pattern, consisting of straight lines, it is also possible to mimic Fig. 6.4.

To make the left-hand bar fringe-free, the added grid should have a slightly smaller fringe spacing than that of the frame, but the difference is so small that in this case the same grid can be used all the time. Figure 6.8 can thus be used to produce all the described fringe patterns. It is, however, not possible to use this moiré pattern method to transform Fig. 6.3 into Fig. 6.8 without ambiguity, the missing information about direction can never be regained by any moiré pattern method.

There is one more way in which Fig. 6.8 can be transformed into something similar to Fig. 6.11, but without having to produce a grid with a particular

Fig. 6.11. Placing a grid of vertical fringes on top of Fig. 6.8 transforms its fringe pattern into that of Fig. 6.3. Thus compensation for rigid body motion can also be performed by a moiré effect in the image plane.

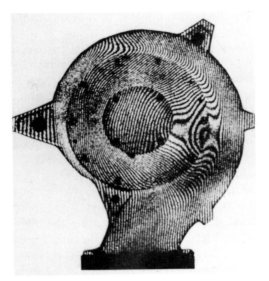

Fig. 6.12. A doubly exposed hologram was made of a centrifugal pump with a height of about 1 m. Between the two exposures the internal pressure was changed. A large rigid body motion, rotation, was also introduced to the pump, producing a great number of almost vertical straight fringes. The desired information concerning the deformation is almost hidden by unwanted fringes.

Fig. 6.13. A grid of straight lines representing the fringes caused by the rigid body motion has been placed over Fig. 6.12. Thus, the moiré fringes represent the difference between the actual pump motion and the rigid body motion. The unwanted fringes have been eliminated and only the required fringes representing deformation are left.

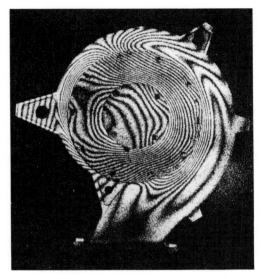

Fig. 6.14. A sandwich hologram was made of the pump of Fig. 6.12. This time the rigid body motion was smaller because no extra rotation was introduced.

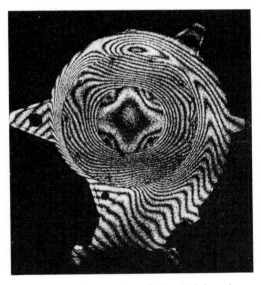

Fig. 6.15. The same reconstruction as that of Fig. 6.12 but the sandwich hologram was tilted until the rim of the centre plate was intersected by as few fringes as possible. In that way the rigid body motion of the centre plate was eliminated and only the required fringes representing its deformation were left. (The centre plate was fixed by four bolts.) Observe the high degree of similarity between these fringes and the moiré fringes of Fig. 6.13.

fringe spacing. However, this special moiré pattern method only works for objects that have at least one axis of symmetry. In such cases one of the bars is symmetric along a vertical axis and therefore the following procedure can be carried out.

Make two transparencies of Fig. 6.8 and turn one of them back to front so that the middle bar is placed beneath its mirror image. The resulting moiré pattern will be identical to that of Fig. 6.3, but with double its number of fringes; this occurs because in this case the deformed object is not compared to its undeformed shape but with itself when deformed in the opposite direction. To study either one of the side bars one transparency has to be translated until the bar under examination coincides with its own mirror image.

6.2.4 Practical example

Let us now examine how moiré pattern effects at the image plane can be applied to a practical example. The object was a centrifugal pump deformed by internal water pressure.

Manipulating the holographic fringes (e.g. by tilting a sandwich hologram) results in a moiré pattern effect between two sets of intersections of hyperboloids and the hologram plate, which is usually situated in the Fourier plane. It is equally possible to manipulate the fringes by producing a moiré pattern effect in the image plane between two sets of intersections of ellipsoids and the surface of the object. If the intention is to investigate the differences between two interference patterns, the two patterns can simply be recorded on two transparencies and then placed one on top of the other. The resulting moiré pattern represents the difference between the two patterns. If one transparency represents rigid body motion of the object while the other represents deformation plus rigid body motion, the moiré pattern will represent deformation alone. If rigid body motion produces approximately straight fringes, one transparency is simply covered by straight lines. For the result to be clear, the rigid body motion should be large compared to the deformation.

A double-exposure hologram was made of a centrifugal pump. Between the two exposures the internal pressure was changed. Figure 6.12 illustrates the result. A large rigid body motion, rotation, has caused a great number of more or less vertical fringes. The desired information concerning the deformation of the centre plate is disguised by these fringes. A grid of straight lines representing the fringes caused by the rigid motion was placed over Fig. 6.12. Thus moiré fringe patterns were formed that represent the differences between the actual pump motion and the rigid body motion (Fig. 6.13). The unwanted fringes were eliminated and only the required fringes representing deformation are left. Figure 6.14 shows a sandwich hologram reconstruction of the centrifugal pump. To study the deformation of the centre plate, the sandwich was tilted during reconstruction until the interference fringes of Fig. 6.15 arose. These are almost identical to the moiré fringes of Fig. 6.13.

This example demonstrates that unwanted fringes can be eliminated in two ways: either by a moiré pattern effect in the plane of the image of the object or by an analogous effect in the plane of the hologram plate. The first method, however, has limitations because the moiré patterns are not coherent and therefore coarse, dark fringes cannot be eliminated. The moiré pattern on the hologram plate does not have this limitation and that is one of the reasons why sandwich holographic interferometry is expected to have great significance in future.

6.3 Studies on a vertical milling machine

The first set of experiments[4] carried out by the author was made directly on the floor of the large laboratory hall ($30\,\text{m} \times 10\,\text{m}$) at the Division of Production Engineering, Royal Institute of Technology, in Stockholm. The test object was a Swedish milling machine of vertical knee type (Sajo type VF54 from Sandén) with a height of about $2\,\text{m}$ and a weight of about $2000\,\text{kg}$. The machine was painted white with ordinary spray paint. It stood directly on the floor (concrete covered by hard fibre-board), which rested directly on the foundation of the building.

The laser used was an ordinary Spectra-Physics He–Ne laser (model 125) with an output of some $60\,\text{mW}$. The exposure time was $10\,\text{s}$ on Agfa-Gevaert Scientia $10\,\text{E}\,75$ plates. The distance from the spatial filter (i.e. the source of the divergent illumination beam) to the machine was some $10\,\text{m}$ and this was also the distance between the machine and the hologram plate.

There were many reasons for the use of these rather large distances. One was to optimize the utilization of the limited coherence length of the laser light, which was only some $150\,\text{mm}$ (see Section 7.4.2). Another reason was to minimize the risk of contour fringes being produced by large rigid body motion (see Section 7.4.2). A third very important reason for a large distance between the object, and the light source and hologram plate (the last two items being in close proximity) is related to the holo-diagram. The more plane and the more equally spaced are the ellipsoids of the holo-diagram, the more constant (and therefore the closer to 0.5λ) is the sensitivity over the object's surface. (The k-value of the holo-diagram was in fact almost constant and was close to unity.) Another way to say the same thing is that the sensitivity vector will have a more constant amplitude and direction over the surface of the object because the angles of illumination and observation will vary less. Thus our arrangement resulted in these angles being approximately $90°$ with respect to the surface of the object, and therefore the sensitivity was

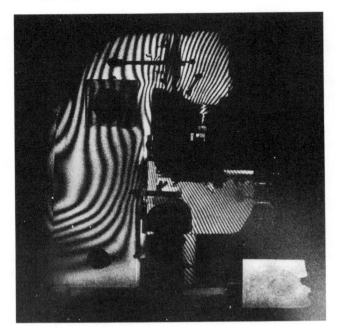

Fig. 6.16. The cutting force of the milling machine was simulated by static loading. Every fringe represents a displacement of about 0.3 μm normal to the plane of the photograph. Straight fringes represent a tilt around an axis parallel to the fringes. Curved fringes represent deformation. A fixed reference surface is seen on the lower right. See also Fig. 7.12.

close to 0.5λ for out-of-plane motion and close to zero for in-plane motion (parallel to the ellipsoids).

On the floor to the right of the machine (see Fig. 6.16) was placed a heavy (25 kg) piece of steel to be used as a fixed reference surface in relation to which the motions of the machine could be defined. As the forces deforming the machine were internal, there was little reason to believe that the floor would transmit any deformations from the machine to this reference surface.

6.3.1 No fringe manipulation

The vertical knee-type milling machine used in the first set of experiments was deformed by a simulated cutting force produced by a pneumatic membrane which was placed between tool and workpiece. A hologram plate with its emulsion forward was placed behind a compensation plate in the holder, and the first exposure was made with the machine at rest. The plate-holder was emptied, and another hologram plate with its emulsion forward was placed in front of the compensation plate. Air pressure was applied to the membrane to simulate a low cutting force, and after a few minutes a second exposure was made. The two hologram plates were processed and put together to form a sandwich.

The result is seen in Fig. 6.16. At the top of the machine is the head, while the large part protruding to the right of the main body is the knee, which is supported from the base by a heavy screw. Above the knee is the table, which has totally disappeared from the holographic image because it reached far outside the ellipsoid representing the limited coherence length (see Fig. 7.12 following). On the floor to the right of the machine is the fixed reference surface against which the motions of the machine could be defined. By tilting the sandwich hologram during reconstruction it was possible to move fringe patterns eliminated from any part of the milling machine to this surface.

Figure 6.16 displays the motions of the machine with reference to the floor, each fringe representing a displacement of $0.32\,\mu m$ along the line of sight. Thus the motion of any part of the machine can be calculated by counting the number of fringes that have to be crossed if a path is traced from the foot of the machine to the point under scrutiny. Areas that have not moved in the line of sight are fringe-free (the foot of the machine; the reference surface). Areas covered by straight parallel fringes have tilted rigidly without deformation. The tilt is larger when the fringes are closer together. It can be seen that the knee has tilted most of all. The angle of tilt can be calculated by counting the number of fringes across its known width from left to right. Areas of the machine that are covered by curved fringes have been deformed. (The main body has been deformed by the moment caused by the force acting between tool and table.)

6.3.2 Compensation for rigid body motion and evaluation of direction of motion

From Fig. 6.16 it is difficult to judge the deformation of the head of the milling machine because its fringe pattern is caused mainly by the large torsion within the main body. After the sandwich hologram had been tilted until the number of fringes crossing the head became as low as possible, it was, however, easy to study the deformation of the head without any disturbances from the rest of the machine (Fig. 6.17). From the angles through which the sandwich hologram had to be tilted to change the picture from that of Fig. 6.16 to that of Fig. 6.17, the angle of tilt of the machine head could be evaluated with respect to magnitude and direction. This result corresponds to what could be found by counting the fringes on the head in Fig. 6.16 or the number of fringes per unit length on the reference surface in Fig. 6.17.

When the same procedure was carried out in order to study the deformation of the knee it was found that only one or two fringes were left (Fig. 6.18).

Fig. 6.17. The sandwich hologram of Fig. 6.16 was tilted so that the deformation of the machine head could be studied without influence on the fringe pattern from the deformation of the whole machine. The fringes that have been removed from the head reappear at the reference surface. When the number of fringes decreased at the head it increased on the knee, indicating that head and knee tilted in different directions.

Thus, the applied force was too low to produce a good indication of the weaknesses of the knee, which is obviously much more stable than the vertical support connecting it to the main body. This attachment appears to be the weakest point of the machine.

From Fig. 6.18 one of the advantages of fringe manipulation can be clearly seen. If a horizontal line is traced at the height at which the knee is attached to the main body, the total number of fringes intersected will be zero on the knee but some 50 on the main body. Thus it can be concluded that the latter number represents the tilt of the main body in relation to a knee that appears to be fixed in space because of an applied sandwich tilt. If the width along a horizontal line at the main body is now assumed to be approximately the same as at the knee, it can be concluded that the knee has tilted in such a way that its outermost corner has moved a distance representing 50 times half the wavelength or some 16 μm. This displacement must be compared to that calculated when the main body appears to be fixed in space, as illustrated in Fig. 6.16, which was made without any sandwich tilt.

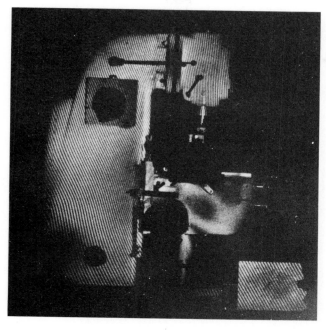

Fig. 6.18. To study the deformation of the knee the sandwich hologram of Fig. 6.16 was tilted in a direction opposite to that of Fig. 6.17. The deformation was, however, very small and resulted in only about one fringe.

If an attempt had been made to carry out the same calculation on Fig. 6.16, it would have resulted, instead, in the number of fringes on the main body being nine and a total of 41 fringes on the knee. The most obvious way to evaluate this information would be to conclude that the entire machine has tilted in one direction so that the relative tilt between the main body and the knee would be $41 - 9 = 32$ fringes. That this result of 32 fringes is erroneous has been proved by the calculation using Fig. 6.18 instead. Thus an error of 36% was, in this practical case, made because the change in direction of the angle of tilt between the knee and the main body was not observed.

To produce larger deformations another pair of holograms were combined which represented a force difference that was three times larger than that already studied. Another change was also made because during the earlier experiment it was found that the sandwich tilt needed to eliminate the fringes on the table was uncomfortable large. By making the separation between the two emulsions of the sandwich hologram larger (d in Fig. 3.23), the angle of tilt could be kept smaller. The two holograms were placed with their emulsions outward instead of forward, and an additional third glass plate (a

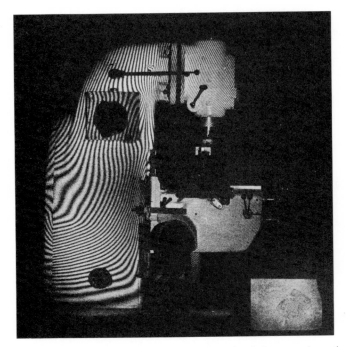

Fig. 6.19. The load of the machine was increased so that it became three times larger than in Fig. 6.16. Without any tilt of the sandwich hologram the fringes on the knee were so closely spaced that they could hardly be counted.

hologram plate with the emulsion removed) was inserted between the two plates to provide extra separation.

Figure 6.19 shows the result before any tilt has been introduced to this new sandwich hologram. The tilt of the knee is now so large that the fringes could not quite be resolved. However, after a tilt of the sandwich hologram during reconstruction, fringes appeared on the knee, and after some further tilt the number of fringes crossing it was set at a minimum (Fig. 6.20). It is now possible to measure the deformation of the knee of the milling machine in spite of this deformation being so small that it caused only about eight of the approximately 125 fringes that originally existed. From the angle at which the sandwich hologram had to be tilted to transform Fig. 6.19 into Fig. 6.20 it was also possible to calculate the magnitude and direction of the angle of tilt of the knee, in spite of the fringes originally being so closely spaced that they could not be accurately counted. The speckles of Fig. 6.20 are larger than those of Fig. 6.19 for the following reason. When the sandwich was tilted, the fringes started to move when the point of observation was changed. Therefore the aperture of the camera that was used to make the photograph had to be so small that the whole area of the lens used appears to see the fringes from

Fig. 6.20. The sandwich hologram of Fig. 6.19 was tilted so that the deformation of the knee could easily be studied. The closely spaced fringes that were moved from the knee to the reference surface could no longer be resolved because of the large speckles. However, the magnitude, direction, and sign of knee tilt could be evaluated from the tilt of the hologram.

the same position. Because of the small aperture the speckles are large in Fig. 6.20. There are effective means available to get around this problem; however, in these experiments the author was satisfied provided that the quality was such that the number of fringes could be counted.

It was, of course, also possible to study the deformation of the head, but then it was necessary to tilt the sandwich hologram in the opposite direction (Fig. 6.21). The fringes on the reference surface could, however, not be counted because of the large speckle size.

In trying to find the limits of the sandwich method the force difference applied to the machine was doubled and a third sandwich hologram was made. This time a thicker extra glass plate was used to separate the hologram plates so that the distance between the emulsions was some 9 mm. The quality of this hologram and the resolution of the fringes were not reduced by this large separation. The result without any sandwich tilt is seen in Fig. 6.22. No fringes are seen on the knee, but it was still possible to eliminate the approximately 250 fringes that had theoretically originally existed there to

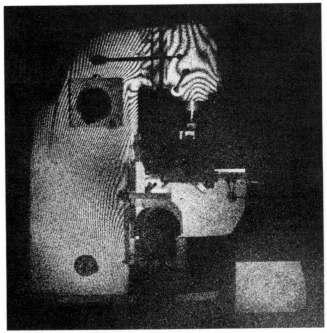

Fig. 6.21. The sandwich hologram of Fig. 6.12 was tilted to make possible a study of the deformation of the machine head.

study the deformation which had resulted in about 20 fringes (Fig. 6.23). In later experiments it has so far proved possible to eliminate up to 500 fringes, and the limit has yet to be reached.

6.3.3 Moiré pattern methods

Finally, one experiment was carried out to study how manipulation using the moiré effect at the image plane corresponded to the results produced by sandwich tilt and described above. A grid of straight lines was projected on Fig. 6.16 by means of an ordinary enlarger. The direction and spacing of the fringes was so adjusted that as few fringes as possible intersected the head of the machine and Fig. 6.24 was produced. There is good agreement between the patterns of Figs 6.24 and 6.17, indicating that the two methods give identical results.

It is also seen that the fringes on the reference surface of Fig. 6.17 are identical to the lines of the added grid, which, of course, they should be. The reason is that the grid represents the pattern that would have formed if the head had not been exposed to any deformation but only to its actual rigid body motion.

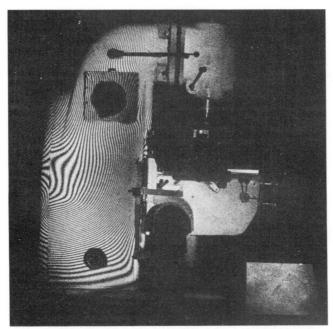

Fig. 6.22. The load was finally increased still more so that the fringes on the knee could not be resolved at all.

Fig. 6.23. After some tilt of the sandwich hologram of Fig. 6.22, fringes appeared on the knee, and further tilt revealed its deformation.

Fig. 6.24. A grid of straight lines placed on top of Fig. 6.16 transformed the fringe pattern on the head to that of Fig. 6.17. The fringe spacing and direction of the added grid was identical to the fringe pattern on the reference surface to the right of Fig. 6.17.

One of the disadvantages of the moiré method is clearly demonstrated in Fig. 6.24. Moiré fringes are seen only on the head; the rest of the machine is without any moiré effect because the difference in angle or spacing of the added grid lines is too large compared to those of the original interference fringes.

6.4 Studies on a horizontal milling machine

For the next set of experiments a plain horizontal knee-type milling machine (Oerlikon model MN2H), was tested at the Department of Machine Tools and Production Engineering at the Swiss Federal Institute of Technology (ETH) in Zürich, Switzerland. The machine was about 1.7 m̄ high and stood directly on the ground floor in a workshop which was only about 7 m × 10 m. Because of the limited space it was all but impossible to get further away from the machine than 4 m, and the spatial filter had to be placed within 0.5 m of a window. ETH is situated in the middle of Zürich, and the stability problems were initially rather severe because of the heavy traffic of lorries, buses, and trams at the street corner outside the laboratory. The floor was covered with cork tiling, a material which is difficult to work on because it retains previous weight depressions and dimensions for a long time after a load has been changed. It was therefore necessary to wait for several hours after changes in the holographic set-up before a hologram could be exposed successfully.

A Spectra-Physics argon laser model 165 was used for illumination; this had an output of 0.5 W in the green 514.4 nm wavelength. After some adjustment of the étalon it could be proved by using a simple interferometric test that the coherence length (see Section 7.1.2) was more than 2 m and therefore did not put any tight restrictions on the holographic configuration. The laser was placed along the wall on an ordinary table, and the beam was deflected through 90° by a mirror and directed on to a spatial filter which was fixed to a very stable, heavy steel table (weighing some 300 kg), on which the plate holder was also fixed. The reference mirror was placed about 0.5 m from the milling machine on a stable stand that weighed around 100 kg. All the heavy equipment used was of the type that can be found in any well equipped heavy duty machine shop.

The milling machine itself had a green gloss finish, and there was an express intention that it should not be repainted or treated in any special way. Therefore it was unavoidable that those parts of the machine that produced a direct reflex got much brighter in the hologram than other parts.

The holograms were exposed at night because then the activity in the Institute was much lower and most of the outside traffic had stopped. The force deforming the machine was produced simply by turning a screw which pressed against the shaft of the machine. The screw was rotated in a threaded hole in a steel spring which was clamped in a vice fixed to the table of the machine. Between the two exposures the author was able simply to go up to the machine and rotate the spanner fixed to the screw. The doubly exposed spanner can be seen throughout Figs 6.25–6.29, in which its positions before and after the application of the force produced a bright inclined V which is seen in front of the horizontal shaft of the machine.

6.4.1 No fringe manipulation

The sandwich holograms were made in the same way as was described earlier. Each hologram plate was exposed with its emulsion forward in combination with a compensating plate in front or behind. No extra glass plate was used to increase the separations of the plates. The distance between the machine and the plate-holder was about 4 m. The exposure was 5 s using Agfa-Gevaert Scientia 10 E 75 plates.

To study the stability and find a suitable deformation range, real-time fringes (Section 3.2) were first observed. These fringes were never stable. During the daytime they were almost impossible to see, and even at night

Fig. 6.25. The deformation of a horizontal milling machine was studied using double-exposure holography. Temperature gradients in the air produced fringes on the reference surface at the lower left.

they never stopped moving. No attempt was made to decrease the influence of air movements by building a tent around the machine, but the cold airflow from the window and the hot air from the radiatior were controlled to some extent by insulation with plastic foam. As usual in sandwich holography a reference surface in the form of a heavy piece of steel was placed close to the machine (lower left-hand corner of Figs 6.25–6.29). After a settling down period of a few hours an ordinary double-exposure hologram was made and it was found that the reference surface was also useful in this type of holography. As seen in Fig. 6.25, it was covered by two fringes that had apparently been caused by air turbulence. The electric cable to the laser, which passed in front of the machine, was under-rated and was so hot that it could not be held by hand. The cable was changed and moved behind the machine, after which these fringes disappeared.

A sandwich hologram of the machine is seen in Fig. 6.26 and, as the fringes are very similar to those of the double-exposure hologram and there are no fringes on the reference surface, it can be concluded that this hologram presents a correct image of the deformation. The fringes in the lower right-hand part of the machine have changed very slightly compared to

Fig. 6.25, probably because of the removal of the cable. There is still a small variation in brightness on the reference surface, indicating an error of a fraction of a fringe and probably resulting from air movement.

All the sandwich holograms photographed during this set of experiments were reconstructed with the green light from an argon laser that had the correct divergence. When an attempt was made to reconstruct these holograms using the red light from a He–Ne laser, fringes of high contrast were formed but these fringes were erroneous because of the difference in wavelength. Thus it was impossible to make the reference surface fringe-free. About two fringes, in the form of concentric circles, could never be removed. The error was so small that it could be neglected when single details were studied which were smaller than half the reference surface, but if it was necessary to compare the relative displacements of parts over the whole machine surface the errors would have to be taken into account. However, using the argon laser for reconstruction, these errors were totally eliminated.

Figure 6.26 shows the milling machine with its main body to the right, on top of which the cross-arm rests with its head at the upper left. Two supports stretch downwards from the cross-arm to the shaft on which the cutter is

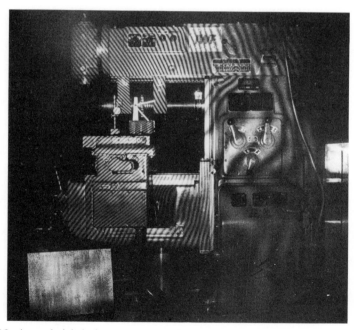

Fig. 6.26. A sandwich hologram was made recording the same situation as in Fig. 6.25, but a heat source in front of the machine had been removed. The deformation and motion of the left support of the shaft were difficult to calculate because the whole cross-arm of the machine had bent and made a torsional motion.

fixed. The vice with the loading screw is clamped to the table, which rests on a saddle that can slide on the knee and which finally is supported by a heavy vertical screw. It is difficult to use Fig. 6.26 to study the deformation of the left-hand support (which is just behind a small dial-gauge) because of the torsion and bending of the entire upper part of the machine. Using simply the photograph in Fig. 6.26 it is not possible to find the direction or amplitude of deflection.

6.4.2 Evaluation of relative motions

Figure 6.27 shows the sandwich hologram of Fig. 6.26 tilted during reconstruction until the upper part of the machine close to the support is fringe-free. By counting the remaining fringes (five) on the support its deflection in relation to the neighbouring part of the machine is calculated $(5 \times 0.26 = 1.3 \, \mu m)$. The direction of the deflection is found by studying the direction of tilt of the sandwich which is needed to reduce this number of fringes. That the total upper part of the machine could not be made

Fig. 6.27. The sandwich hologram of Fig. 6.26 was tilted so that the cross-arm became fringe-free around where the left-hand support is fixed. The number of fringes on the left-hand support then became a measure of its motion in relation to the cross-arm.

Fig. 6.28. The sandwich hologram of Fig. 6.26 was tilted so that the fringes at the left-hand support were rotated through 90° compared to those of Fig. 6.27. That way the tilt of the support is indicated by the sharp change in angle as the fringes pass from the support to the cross-arm. The curvature of the fringes on the support indicates its deformation, from which stress and strains can be calculated.

fringe-free in Fig. 6.27 proves that it has been deformed by torsion, which is natural considering the way in which the force was applied. The right-hand support has almost no fringes at all, which proves that it was almost undeflected.

Even from Figure 6.27, however, it is difficult to say exactly how the deformation of the left-hand support was distributed over its entire length. Thus, by tilting the sandwich hologram, the fringes were rotated through close to 90° so that the fringe inclination became a measure of the tilt of the support and the curvature of the fringes a measure of the tilt gradient or the bending of the support (cf. Fig. 6.8). The result can be studied in Fig. 6.28, in which it can be seen that the support was bent in the form of an S. This bending is, however, slightly less than one fringe, e.g. less than 0.26 μm. Thus the main reason for the total deflection of 1.3 μm was a tilt of the support because it was badly fixed to the machine at its upper end. This fact is further revealed by the abrupt change of fringe angle where the fringes pass from the support to the machine.

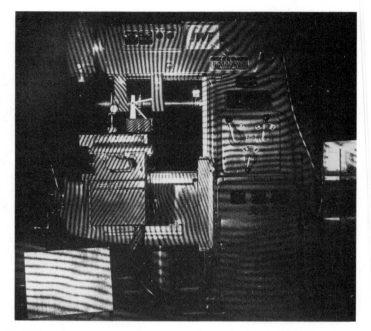

Fig. 6.29. When the same procedure was performed with the right-hand support, the straight vertical fringes indicate no tilt and no deformation.

When the same procedure was repeated for the right-hand support (Fig. 6.29) it was found that it had been neither tilted nor deformed. Apparently the entire load had passed through the left-hand support. Using the methods described here, together with the information from a single sandwich hologram, it is possible to inspect every detail of the entire machine.

6.4.3 Reading fractional fringes

It is now time to be a little more sceptical. Can one really trust a sandwich hologram measurement that is of less than one fringe, as on the left-hand support of Fig. 6.28? The author believes that generally this would be unrealistic because of the practical limitations. There are the influences from air turbulence and temperature gradients and also from variations in the thickness of the glass of the hologram plates.

In the first hologram of this set of experiments (Fig. 6.25) there were erroneous fringes at the reference surface with such a small spacing that they would produce about two residual fringes at the support. In Fig. 6.26 these residual fringes, believed to have been caused by a temperature gradient in the air, have been eliminated, but there remains a variation in brightness at the reference surface which could indicate a fraction of a fringe. This effect is still more clearly seen in Fig. 6.29, where the fringes are out of alignment by about a quarter of a fringe over the length of the reference surface. The support has about a third of that length, so that the corresponding error should therefore amount to something like a twelfth of a fringe. As the temperature gradient usually decreases with the distance from the floor, there are good reasons to believe that the error caused by this effect is still smaller than the calculated value.

Thus, even if the temperature of the air has no significant influence on the fringe pattern, there still remain the errors caused by the sandwich hologram itself, e.g. by variations in glass thicknesses. The simplest check of these errors is to move the observing eye along the hologram plate and to study the fringes on the reconstructed object. As this process caused a fringe motion but no change in fringe curvature at the support it was concluded that the fringes reveal the true deformation and nothing else.

Finally a test was carried out using the moiré pattern method. A grid of straight lines was projected on top of Fig. 6.25 or Fig. 6.26. The result is seen

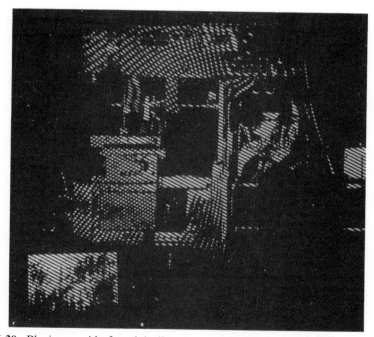

Fig. 6.30. Placing a grid of straight lines on top of Fig. 6.26 produces moiré fringes on the left-hand support that are similar to those of Fig. 6.28. The angle and spacing of the added fringes are identical to the fringe pattern on the reference surface of Fig. 6.28.

in Fig. 6.30. At a particular frequency and angle a moiré pattern formed that was almost identical to that of Fig. 6.28. There is, however, a small discrepancy because the fringe frequency was not absolutely correct, which can be seen by counting the fringes at the reference surface. Apart from this slight difference there is still the same S-shape of the fringes at the support. As the double-exposure hologram together with the grid produces the same result as the sandwich hologram, it is concluded that the errors of the latter are negligible. Thus the result of Fig. 6.28, i.e. that the support being examined was deformed in the shape of an "S" in very much the same way as the middle bar of Fig. 6.8, can be trusted.

6.4.4 Vibrations

Finally, vibration analysis was carried out using conventional time-averaged holography (Section 3.4). An electromagnetic vibration generator (Goodmans type V50 Mk 1) was clamped to the table of the milling machine and attached to the shaft. The direction of the forced vibrations was the same as that of the static load during the earlier experiments. To find the resonance modes of the

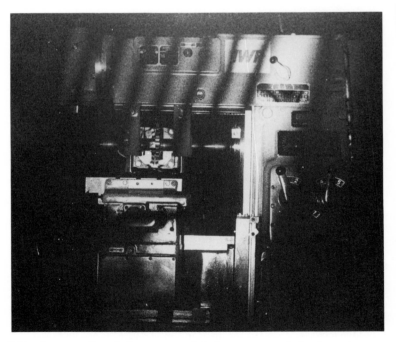

Fig. 6.31. A conventional time-averaged hologram recorded when the arm was excited at its lowest resonance mode. The frequency was 75 Hz and the fringes indicate a simple bending motion, the node being the bright, broad, almost vertical band.

machine an electric vibration transducer was fixed to its upper part. The lower resonance frequencies that had sufficient amplitudes were noted and a time-averaged hologram was made at each resonance frequency.

Surprisingly no image whatsoever was observed during initial trials. After about 1 h of experimentation a first low quality hologram was finally produced, and after some further time the quality was improved. It was concluded that the vibrations forced the machine to start moving on its foundation until it settled down in a new equilibrium position. Another observation was that to get a high quality image a higher ratio between object and reference beams was needed when the time-averaged holograms were made. No measurements were made but simply by moving the reference mirror further away from the centre of the illuminating beam it seems probable that the ratio would be increased from $1:10$ to $1:2$.

Figure 6.31 shows how the upper part of the machine vibrates at its lowest resonant frequency (75 Hz). The broad bright fringe which is slightly inclined to the vertical represents the nodal line. The upper part of the machine simply vibrates forward and backward as if it rotates around the nodal line. The maximum amplitude at its left-hand end was $7 \times 0.25\lambda = 0.9\,\mu m$. The

Fig. 6.32. The same situation as in Fig. 6.31, but a higher resonance mode was excited. The frequency was 125 Hz, and the fringes indicate a mainly torsional motion of the cross-arm. Vibrations of the shaft and the data plate (upper right) are also seen.

amplitude of vibration of the lower parts of the machine was too small to be detected.

The second lowest resonance mode (325 Hz) is shown in Fig. 6.32. If the top part of the machine had vibrated in a single torsional mode around a horizontal axis, there would have been a long bright horizontal fringe. The nodal island indicates a vibration mode made up mainly as a result of torsion but also of a slight bending. The two supports that go down to the shaft are also fringe-covered, so that it can be seen that the shaft close to the left support was vibrating with an amplitude of $9 \times 0.25\lambda = 1.16\,\mu\text{m}$. The amplitude of vibration of the main body and the knee was too small to be detected, but it is possible to see vibration patterns on the data plate and on the cable to the right of the machine.

During studies of the milling machine in Stockholm some holographic measurements were also made during the actual cutting process. Because of the random motions and vibrations produced by the working machine it was necessary to use a double-pulsed Q-switched laser with a pulse separation of about 1 ms. The resulting patterns in the holographic image were very similar to those obtained by using a static load. The patterns were, however, difficult to evaluate because the sign of the displacement was not known and could not be found from the conventional double-exposure hologram being used. Therefore, it was decided to develop a sandwich hologram method that could be used even for double-pulsed holography with a pulse separation in the millisecond range. The results of this method are described in the following section.

6.5 Studies on a hand-held drilling machine at work

If the unique possibilities of double-pulsed hologram interferometry are to be fully utilized it is not enough to be able to freeze the motion during each exposure by using very short illumination pulses. A method is also required to eliminate the influence of the object's rigid body velocity.[5,6] Later on it will be seen to be of great importance to be able to find the directions of the velocities being examined. In dynamic events this direction is not usually known in advance, in contrast to the situation which is often the case with displacements caused by static loading. Ordinary vibration patterns formed by time-averaged or by double-pulsed hologram interferometry do not reveal the phase relationships between different vibrating surfaces. The reason is that the direction of motion (outwards or inwards) is missing in the conventional holographic process. The possibilities for solving these problems in a practical way were initially studied by applying the sandwich hologram method to make measurements of the vibrations of a hand-held drilling machine at work. To be able to study the dynamic behaviour of such an unstable object it was necessary to combine a double-pulsed ruby laser with a spinning hologram holder.

The earlier technique for sandwich holography utilizing a continuous laser can, of course, also be used for pulsed holography, provided that the pulses are separated by a sufficient time that the plates can be changed between pulses. However, the main purpose of the technique described in this section is to make fringe manipulation possible even when the delay time between the two pulses is less than 1 ms and the plates cannot easily be changed. The approach is to expose a rotating pair of plates so that the images from the two pulses are separated at the sandwich plates and then to combine the recording on the rear plate from the first pulse with the recording on the front plate from the second pulse. To separate the images from the two pulses a relatively high speed of the plates is needed.

Thus two ordinary hologram plates were placed, one in front of the other, at a disc which was rotated by an electric motor (Fig. 6.33). In front of the plates was a metal screen with a small slit. During the first laser pulse the illuminating light could reach the plates only at the area behind the slit. Between the pulses, which had a separation of less than 1 ms, the plates had time to rotate sufficiently for new areas to be exposed behind the slit. The recording from the second pulse was thus separated from the first.

After processing the plates were rotated, one in relation to the other, in such a way that the first exposure of the rear plate coincided with the second

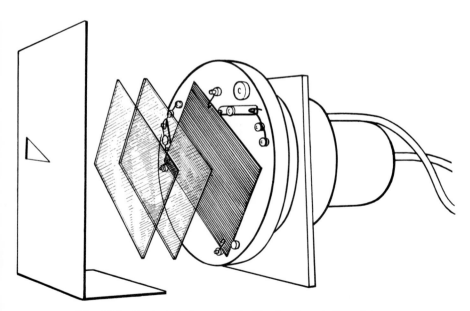

Fig. 6.33. Exploded view of the holder for the rotating plates.

exposure of the front plate. The two plates were then glued together and reconstructed just like an ordinary sandwich hologram, the only difference being that the viewing angle was considerably limited because the utilized areas of the plates were only about $1 \, \text{cm}^2$.

6.5.1 Introductory experiment with a rotating cylinder

The experimental set-up consists of a rotating, white-painted cylinder ($\phi = 110 \, \text{mm}$ and $n = 0.75 \, \text{r.p.m.}$), a fixed reference surface (also painted white) and a mirror to direct the reference beam to the plates (Fig. 6.34). The light from a ruby laser (Holobeam 651) was spread by a diverging lens to illuminate the objects. The rotating plate-holder was placed in front of the object. The plates used for the experiments were without any antihalo coating so that the rear plate could be exposed through the front plate. No triggering is needed: the rotating plates can be exposed at any time. For the results shown the plates were exposed at $n = 6800 \, \text{r.p.m.}$, and the separation time between the two pulses from the laser was $600 \, \mu\text{s}$.

The separation between the two spots at the plates was 16 mm at a distance of 45 mm from the axis of rotation. The quality of the holographic images at the rotation speed of 6800 r.p.m. was as good as if the plates had been exposed at rest.

The evaluation equipment was designed to make it possible to combine the image from the first pulse with the image from the second pulse and to bond the plates together. Therefore a holder was made similar to the one used for exposing the plates. The positions of the pins, ball bearings, and clamping devices were exactly the same; however, there was one holder for each plate, so that the plates could be rotated in relation to each other around the same axis as the axis of rotation during exposure.

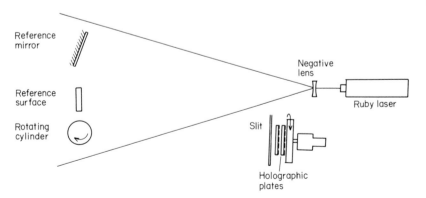

Fig. 6.34. Experimental set-up.

Fig. 6.35. Reconstructed sandwich hologram with fringe-free reference surface.

After positioning the developed plates in the holder, a coarse rotation of one plate was made to align the first and second images. Then, by observing the holographic image and using a micro-positioning device for fine adjustment of rotation, the fringes appeared. The adjustments were completed when the reference surface was fringe-free; after that the plates were glued together. The bonded plates were removed from the holder, and the plates could be evaluated by hand as for an ordinary sandwich hologram.

The procedure for fitting the plates in the evaluation holder was to rotate them for interference between the two pulses and glue them. This only took about 5 min for each pair.

When the sandwich hologram was observed, vertical and parallel fringes appeared on the cylinder because of its rotation. Figure 6.35 shows a photograph of the pulsed sandwich hologram. These fringes could be eliminated by tilting the sandwich around a vertical axis in the same direction as the rotation of the cylinder (Fig. 6.36). By introducing a known horizontal tilt of the sandwich the inclined fringes at the cylinder also give information on the direction of rotation (Fig. 6.37). The limited size of the hologram sometimes makes it better to project an image on a screen for evaluation.

Because the reconstruction is made by means of a He–Ne laser an error is introduced in the fringe pattern as a result of the difference in wavelength between the He–Ne laser and the ruby laser. As is seen in Fig. 6.35 the fringes at the cylinder are slightly bent. Therefore, when using this technique the

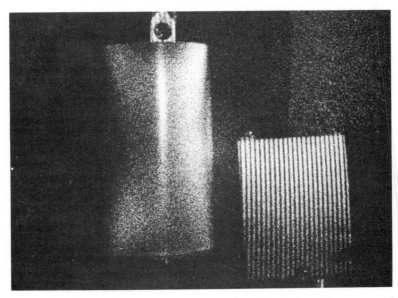

Fig. 6.36. Fringe-free reconstruction of the rotating cylinder from the same sandwich hologram as in Fig. 6.35. The hologram is tilted in the same direction as the object was rotated.

Fig. 6.37. Reconstruction with horizontal fringes on the reference surface. The inclined fringes at the cylinder also give information on the direction of rotation.

evaluation should be made using the same wavelength as during exposure if extreme accuracy over a large viewing angle is needed. To fulfil this requirement is not too difficult, there exist pulsed lasers with a repetition rate of 50 Hz. At the evaluation stage it should also be possible to use a dye laser operating at the same wavelength as the pulsed laser.

6.5.2 Conditions during the main experiment

The experimental set-up (Fig. 6.38) consisted of a drilling machine (C) held by hand and operating on an iron plate, a fixed reference surface painted white, and a mirror (D) to direct the reference beam to the plates in the rotating plate-holder (B). The light from a ruby laser (Holobeam 651) was expanded by a diverging lens (A) (of focal length 6.5 mm) to illuminate the objects.

The plates were developed and placed in the evaluation equipment. After positioning the plates in the holder, a coarse rotation of one plate was made to align the first and second images. Then, by observing the holographic image and using the micro-positioning device for fine adjustment of rotation, it was possible to find a position at which the fringes started to appear. The adjustment was finished when the reference surface was fringe-free. After that the plates were glued together.

The photographs of the pulsed sandwich hologram were made in such a way that a laser beam was passed through the sandwich hologram and reconstructed a real image which could be projected on to a screen and then photographed. This procedure was more convenient because of the limited size of the exposed areas of the plates. The fringe patterns in the projected image were always identical to those seen in the virtual image (some 20 sandwich holograms have been tested in this respect).

The plates used for the experiments (Agfa-Gevaert Holotest 10 E 75 NoAh)

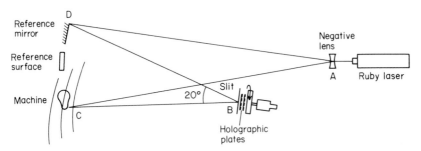

Fig. 6.38. A negative lens at A diverges the light from a ruby laser so that it illuminates the object C, the reference mirror D, and a reference surface. The spinning sandwich hologram plates are positioned at B. The curved lines around the object are parts of the ellipses of the holo-diagram with their common focal points at A and B.

were without any antihalo coating, so the rear plate could be exposed through the front plate. The rotating plates could be exposed at any time, as no triggering device is needed in the system.

The experiments were performed under realistic workshop conditions. No vibration isolation tables were used. However, a suitable darkroom illumination was used in the room.

For the results given in this chapter the plates were exposed at a rotation speed of 6800 r.p.m.; the time separation between the two pulses was 520 µs. The separation between the two exposed areas of the plates was 15 mm, at a distance of 45 mm from the axis of rotation.

Because the reconstruction is made by means of a He–Ne laser, an error is introduced in the fringes as a result of the difference in wavelength between the He–Ne laser and the ruby laser. The error in this case is not more than one extra fringe over the whole height of the drill.

6.5.3 Compensation for rigid body motion

The advantages gained by using static sandwich hologram interferometry have already been described. Dynamic sandwich hologram interferometry, by using, for example, double-pulsed ruby laser light, expands the fringe manipulation method to include the time domain. Thus, it makes possible the elimination of unwanted rigid body velocity and the evaluation of the direction and magnitude of an object's rotational velocity by an analogous rotation of the hologram plates during reconstruction. By rotating the interference fringes through 90°, the fringe spacing is transformed into fringe inclination, which simplifies the measurement of small rotational velocities representing only a fraction of a fringe. The gradient or derivative of the fringe spacing is simultaneously transformed into fringe curvature, from which the stresses and strains caused by different angular velocities can be calculated. Finally, the dynamic sandwich method makes possible the elimination of a large number of fringes, therefore larger velocities can be measured. In studying the hand-held drilling machine emphasis was placed on the different methods available for finding the phase relationships between different vibrating areas.

The photographs shown in Figs 6.39, 6.40, 6.43, 6.44 and 6.47 are all taken from reconstructions of a single sandwich hologram. Let us start by studying Fig. 6.39. The drill is directed downwards and held with two hands, which are partially visible at the top left of the photograph. The index finger of the right hand is almost vertical. A stable bright steel plate, which is fixed to the right of the machine, functions as a reference surface. The background was also stable enough to be used as a reference surface. The interference fringes on the hand-held machine are caused by a mixture of simultaneous rigid body motions and vibrations. In the following it will be shown how it is possible to

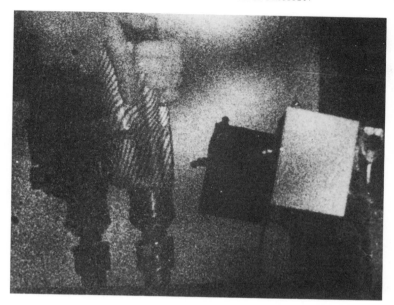

Fig. 6.39. Photograph of the reconstructed image from a rotated sandwich hologram. The hand-held drilling machine was recorded during actual working conditions. The fringes are caused by a mixture of simultaneous rigid motions and local vibrations. The fringe-free reference surface indicates that in this position the sandwich hologram is identical to an ordinary double-exposure hologram.

eliminate the influence of the rigid body motions and concentrate observation on the local vibrations.

Each dark fringe on the reconstructed image of the object, as seen in Fig. 6.39, represents an object point C (see Fig. 6.38) that, between the two exposures, has intersected an odd number of ellipsoids on the holo-diagram (Fig. 1.24(a)). All these ellipsoids have one focus at the light source (A) (the diverging lens) and the other at the point of observation (B) (the hologram plate). Two adjacent ellipsoids represent a path length difference for the light from A to B via C of one wavelength. Thus, a diagram like that of Fig. 1.24(a) can be used to explain how the fringes are formed.

However, this diagram is rather complicated to use because neither the ellipsoids nor the machine surfaces are flat. It would simplify the problem if the ellipsoidal surfaces were flat and equidistant because then the three-dimensional shape of the drill would have no influence on the explantion of the fringe formation. In that case, therefore, the machine can be regarded as being totally flat.

When the hologram was made the distances AC and BC were more than 1 m, but the distance AB was only a few centimetres and the area under

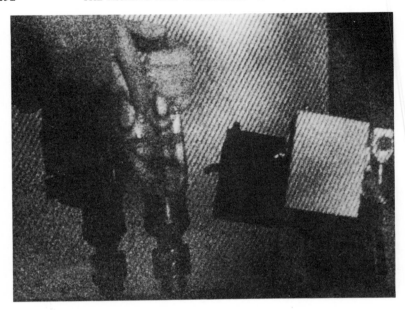

Fig. 6.40. The sandwich hologram of Fig. 6.39 has been tilted so that the number of fringes at the area under study, around the top to the index finger, is as low as possible. The eliminated fringes are seen at the reference surface and in the background. Apparently the rigid motion has been eliminated, so two vibrating areas with a node at the index finger are revealed.

examination at C was but a few square centimetres. As a result the ellipsoidal surfaces (imaginary interference surfaces) could well be approximated by flat equidistant planes; thus, the sensitivity to fringe formation would be constant over the whole surface of the object and would be independent of its three-dimensional shape. Thus it is a simple matter to draw the cross-section of the drill using straight lines.

After these approximations have been made it would be useful to tilt the imaginary interference surfaces so they become parallel to the assumed flat surface of the object. It would then be easy to demonstrate how a curving of the surface of the object causes interference fringes. In Section 6.5 it was indicated how a tilt (ϕ_1) of the sandwich hologram results in a much smaller tilt (ϕ_2) of the imaginary interference surfaces (ϕ_1 and ϕ_2 refer to Fig. 3.22). Thus, it is possible to transform Fig. 6.39 into Fig. 6.40 by tilting the sandwich hologram during reconstruction until the curved fringes close to the index finger break up into two circular fringe systems, indicating two points of maximal vibration amplitude separated by a nodal line at the index finger. Figure 6.41 explains how the centres of the resulting ring systems represent the two points where the imaginary interference surfaces are tangential to the

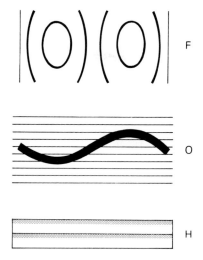

Fig. 6.41. The pattern of Fig. 6.40 is explained as caused by a flat object O vibrating in the form of an S and intersected by straight imaginary interference planes. The intersections cause the fringe pattern (F) as seen through the sandwich hologram (H). The tangent points determine the centres of the two oval fringes. To simplify the picture the sandwich hologram is drawn without tilt when the object is also without tilt.

originally assumed flat surface of the object which has been curved by vibration. In Fig. 6.40 it is easy to see the amplitude and the mode of vibration, but that is almost impossible in Fig. 6.39. However, it is still not possible to say whether the centres of the two ring systems move outwards or inwards, or even whether they move in the same direction.

6.5.4 Evaluation of vibrational phase relations

From Figs 6.41, 6.42 and 6.45 it can be seen how the tilting of the imaginary interference surfaces, caused by tilting the sandwich hologram, results in a change of the separation of the two tangent points representing the centres of the circular fringe systems. In Fig. 6.42 the separation between the tangent points has increased because of a clockwise rotation of the sandwich hologram. Figure 6.43 demonstrates the practical result, in which the centres of the circular fringes have almost moved outside the machine. Figure 6.44 shows how those centres moved very close together when the sandwich hologram was rotated in an anticlockwise direction. Figure 6.45 explains this last result.

Finally, the sandwich hologram was rotated around a horizontal axis so that its top was tilted away from the object. In that case the fringe pattern

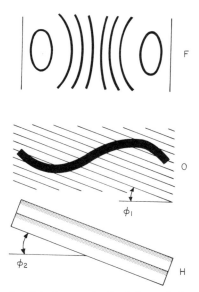

Fig. 6.42. When the sandwich hologram is tilted through the angle ϕ_2, the imaginary interference surfaces are tilted at the angle ϕ_1. In the diagram ϕ_1 is greatly exaggerated, but so is the amplitude of vibration of the object. In reality ϕ_1 should only be a fraction of a milliradian, while ϕ_2 may be some $10°$. It is clearly seen that the centres of the round fringes have moved outwards compared with the situation in Fig. 6.41.

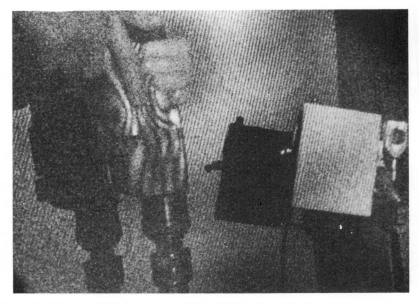

Fig. 6.43. The sandwich hologram of Fig. 6.39 has been tilted around a vertical axis so that its position corresponds to that of Fig. 6.42. The fringe pattern has changed compared with that of Fig. 6.39 in exactly the way predicted by the diagram.

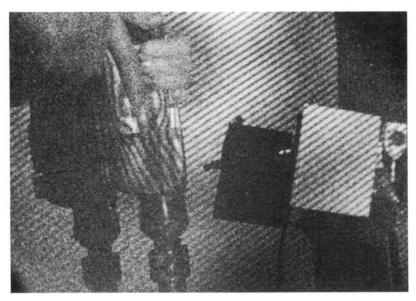

Fig. 6.44. When the sandwich hologram was rotated anticlockwise, the centres of the two oval fringe systems moved closer together until they became hidden by the finger.

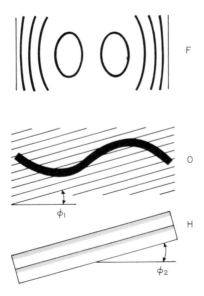

Fig. 6.45. Diagram describing how the fringe pattern of Fig. 6.44 was formed.

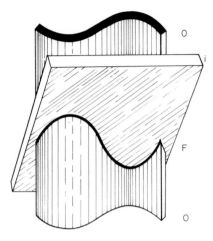

Fig. 6.46. The sandwich hologram was finally tilted around a horizontal axis, which caused one of the flat, imaginary interference surfaces to tilt in the same direction. The simplified object surface OO has been approximated as being bent along only one dimension. The intersection between the curved object surface and the inclined interference plane represents one interference fringe (F).

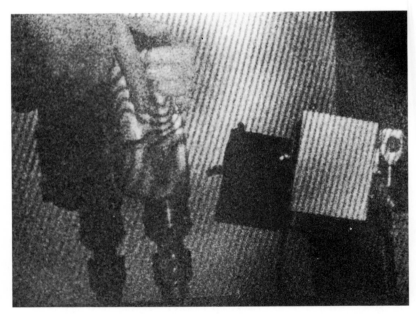

Fig. 6.47. Photograph of the actual fringe pattern caused by tilting the top of the sandwich hologram of Fig. 6.40 away from the object, as described in Fig. 6.46.

(Fig. 6.46) directly represents the vibration modes of the object, because the fringes are formed on the curved surface of the object where it is intersected by the inclined and imaginary interference surfaces. The highest points of the fringes represent maximal motion outwards, and the lowest points represent maximal motion inwards. Figure 6.47 verifies the results of Figs 6.40–6.45, i.e. that the left-hand part of the machine was moving outwards while the right-hand part was moving inwards between the first and second exposure. (This last experiment is closely related to that illustrated by Fig. 6.8, and the equations (6.7) and (6.8) apply.)

There is yet one more way to find the phase relationships from a sandwich hologram of a vibrating object. During the rotation of the sandwich hologram around a vertical axis the fringes move around on the reconstructed image. New circular fringes are formed and move outwards from one of the centres, while the circular fringes at the other centre move inwards and disappear. Even this phenomenon can be explained in a simple way by using imaginary interference surfaces, as shown in Fig. 6.48. When the hologram is rotated in the direction ϕ_2, the interference surfaces rotate in the same direction but with a smaller angular velocity, ϕ_1. The axis of rotation will be in the direction of the source of the reconstruction beam (which was to the right of the object, Fig. 6.36); therefore, the interference surfaces will move away from

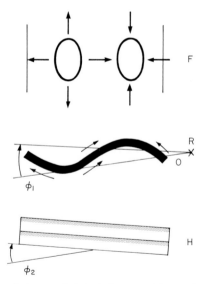

Fig. 6.48. A rotation of the sandwich hologram Fig. 6.40 in the direction of ϕ_1 around any vertical axis caused one of the imaginary interference planes to rotate with angular velocity ω_1 around a vertical axis (R) in front of the source of the reconstruction beam. Thus, during the rotation, the oval interference fringes (F) will expand around the left centre and contract around the right centre.

the observer and thus the fringes expand on the surface of a convex object (bending outwards) and contract on the surface of a concave one (bending inwards).

This method of finding the phase relationship gave the same result as the other methods, but it has the advantage that it is about ten times more sensitive to sandwich tilt. However, as yet no attempt has been made to find out whether this simple use of the fringe motion to explain the direction of the movement of the object can also be used to produce a true quantitative value.

In conclusion, this experiment showed that sandwich hologram interferometry can, in a practical way, be applied to the study of dynamic events such as vibration. The influence of rigid body motions of the object can be eliminated; the phase relationships (motions forwards or backwards) can be found in three different ways, all of which are based on the following simple idea. When the sensitivity to fringe formation is constant over the object, the fringes can be thought of as produced by an originally flat surface which has been deformed by the change in shape between the two exposures and which is intersected by flat, equidistant, imaginary interference surfaces. When the sandwich hologram is rotated during reconstruction, these surfaces rotate in the same direction—their axes of rotation being parallel to those of the hologram plates and situated in the direction of the source of the reconstruction beam.

6.6 Fringe counting rules

Now that the advantages of a well planned fringe manipulation have been demonstrated, one might ask whether these possibilities could also be used in conventional double-exposure holography. The lost information concerning the direction of displacement can usually not be retrieved because the hologram plate does not remember which of the two exposures was first. One exception, however, is if this information about the order of the exposures were included in the recording, e.g. by introducing a known change in the position of the object or an optical component between the two exposures.

In such a case it should, at least theoretically, be possible to separate the two recordings and thus to find the time sequence and consequently the direction of the motion.

6.6.1 Utilizing a large rigid body motion

If the object has been given a rigid body motion that has caused more fringes than the deformation under study, then the relative direction of the two

Fig. 6.49. Sometimes the fringe pattern becomes extremely complicated, so that the counting of the fringes is possible only when following strict rules such as, for example, that counting along a closed loop always results in zero.

motions is preserved. When the direction of the rigid body motion is known then the required motion is found as an increase or a decrease in the fringe frequency. Figure 6.30 is a practical example of how the direction of deflection is found by a moiré pattern effect in such a way that the shape of the deformation is revealed at one of the supports of the spindle axis. The sharp angle of the moiré fringes at the top of the support discloses a discontinuity, a tilt, while the S-shaped fringes show the deformation of the support. From the angle of inclination the conclusion can be drawn that the angles caused by deformation are smaller than those of the tilt; therefore the direction of the deformation, as compared to the direction of the tilt, is disclosed at every part of the support.

Figure 6.49 shows an example of the difficulties sometimes encountered when applying double-exposure holography to complicated mechanical constructions. The object is a "PRESSDUCTOR",† a device that measures large mechanical forces by magnetostriction. The force to be measured is transmitted through the central vertical column, in the middle of which is placed the sensitive element. Two outer branches constitute steel arms that support four flat springs which guarantee motion parallel to the central column.

† PRESSDUCTOR is a registered trade mark of ASEA, Sweden.

When the hologram was made the angles of illumination and observation were almost 90° with regard to the object surface, so that the fringes seen in Fig. 6.49 represent almost purely out-of-plane deformation (bulging).

The object had, however, also made a large vertical rigid body displacement between the two exposures. The reason for this unwanted motion was that the PRESSDUCTOR was tested in a vertical press and its force of some 20 tonnes deformed the mounting table. This downward motion caused roughly two horizontal fringes that moved across the object as the eye was moved up and down along the hologram plate (see Section 5.2.1).

However, these fringes did not move across the object independently of the fringes caused by the deformation. Instead they were added to, or subtracted from, the latter fringes, so that the result was that the whole fringe system appeared to be in motion and the circular fringes started moving inwards or outwards. In other words the fringes moved in different directions at different parts of the object, indicating that the direction of motion could perhaps be found by the use of this technique.

6.6.2 The sign of a fringe

Let us examine the experiment illustrated in Fig. 6.49 more closely. The illumination beam was pointing slightly upwards, so that its path length would be shortened if the object moved downwards. The lower the point is from which the observation is made, the greater will be this shortening of the path length. Thus a motion of the eye downwards during the reconstruction results in each fringe moving further down a deformation valley in order to keep the path length difference between each fringe constant.

As a consequence the fringes move downwards when the observation point is moved downwards, this means that new fringes are formed at the top of the peaks and that the circular fringes are widened at the peaks, while the opposite is the case in the bottoms of the valley where the rings decrease in radius and finally disappear. From this description it can be seen that the situation is very similar to that explained in Fig. 6.48. However, in contrast to the case described, the direction of the large rigid body motion is usually unknown and therefore the true directions of the deformations are still not known. This fact is, of course, a disadvantage, but one does at least know where there are *changes* in direction, which sometimes is of great importance.

It is almost impossible to evaluate the holographic image photographed in Fig. 6.49 without the rules now derived. Assume that the out-of-plane displacements are to be measured along a vertical line at the central column. It is, of course, of great importance not to count the same fringe twice. If a number of circular interference fringes are to be counted, and if one ring is crossed twice, it should be counted as positive the first time and negative the second time (see also Section 6.6.3). Often, of course, the entire interference ring cannot be seen, but only short pieces. In such a case it might be impossible to count the fringes correctly.

However, if during reconstruction the fringes move as the eye is moved along the hologram plate, then the following rules apply:

(1) Check which motion of the eye behind the plate produces the maximal fringe motion (in Fig. 6.49 this is motion along a vertical line).

(2) Choose one direction along this line to be designated as positive (e.g. upwards).

(3) Trace the movement of an imaginary point in the positive direction along the object surface from a fixed point to the point the motion of which is wanted. Count those interference fringes as positive that *meet* the point when the eye is moving in the positive direction.

(4) Count those interference fringes as negative that *overtake* the point when the eye is moving in the positive direction.

In this way it is possible to count the fringes so that the correct number is found even if the sign of this number is still unknown. If this result is not sufficient the author recommends using real-time holography or sandwich holography which both produce a larger amount of information in a more general way.

Sometimes it is possible to check the correctness of fringe counting by counting different paths along the object: the number of interference fringes should always be the same. Thus, tracing an imaginary straight path from the bottom to the top of the central column of Fig. 6.49 should give the same result as a path along either of the side arms. Another check is that tracing a path along a closed loop always should result in the total number of fringes adding up to zero. These two statements depend on the assumption that the object surface and its deformation are continuous. If there are discontinuities then fringes might get lost and thus counting could produce a false value.

One way to get around this problem is by building a bridge over the discontinuity. This bridge could consist of a steel spring that is mounted on the object before the experiment and is fixed to either side of the discontinuity. The number of "lost" fringes can then be found by comparison with the number of fringes formed on the spring.

6.6.3 Varying sensitivity

Previously in this chapter analysis has been restricted to those cases in which the object is so small and far away from the hologram plate that the variations in the angles of illumination and observation can be neglected. Let us now see what differences there will be if those angles change over the surface of the object, resulting in a variation in the sensitivity of the fringe-forming process. In such cases fringe counting should be performed in the following way.

Position the object in the holo-diagram of Fig. 4.19 so that the angles of illumination and observation correspond to those of the actual recording and reconstruction steps. Count the number of dark fringes between a fixed (i.e. undisplaced) point on the object and the point whose displacement is to be evaluated. Multiply this number by the k-value of the object. This gives the displacement in terms of the number of half-wavelengths. The displacement so evaluated is the projection of the real displacement on to the normal to the surface of the ellipse that passes through the object. One way to find this normal is to look for the intersection of the k-circle through the object and the negative y-axis. The normal passes through this intersection.

Assume that the object is so large that a stationary point on it has one value of k and that the point whose displacement is to be evaluated is positioned at a quite different k-value. Which k should be used? The following reasoning and experiments indicate that only the latter k-value should be used.

Imagine that no fringes are formed on the object during the deformation. What really happens is that every point that is displaced causes one cycle of brightness and darkness every time the path length of light from A to B (see Fig. 4.19) is changed by one wavelength. The total number of cycles which any point has experienced can be found because this number is contained in the number of fringes between the deformation point and a fixed point. The number of cycles depends on the continuity of the object's surface since the fringes move along the object in the direction of the fixed point much like pearls on a string. The number of cycles is thus a function only of the deformation of the point under study and the k-value at this point. An example of this is shown in Fig. 6.50.

The object is a 2 m long steel bar (Fig. 6.50). It is positioned between A and B and just underneath the x-axis in Fig. 4.1. The k-value in the middle of the bar is about 11, and at the far end is about 7 (Fig. 6.51). The hologram was exposed once. Then the bar was given a minute rotation around its long axis and a second exposure was made. The bar was not bent during this process. The surface down the long axis of the bar has undergone no displacement and therefore is bright. Displacement increases with distance from the axis, and the first dark fringe is formed at the point where the intersection between the bar and the half-wavelength ellipsoids has moved from one ellipse to the next. The next fringe forms where the intersection has moved three ellipses, etc. The distance separating the fringes is greatest in the middle of the bar, where the k-value is greatest.

The displacement of any point on the surface of the bar can be evaluated as follows. Starting at any point along the long axis (e.g. C of Fig. 6.51) any continuous path can be traced to the point that is to be studied (e.g. D), counting the number of fringes (six) that are crossed on the way. The

Fig. 6.50. Double-exposure holographic image of a steel bar that, between the two exposures of the hologram, has been given a slight rotation around its long axis.

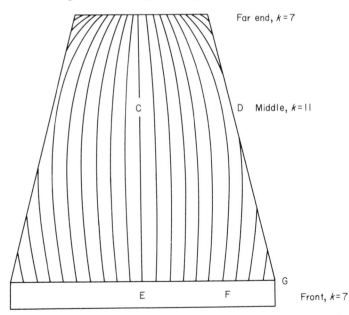

Far end, $k=7$

C   D Middle, $k=11$

E   F   G

Front, $k=7$

Fig. 6.51. The fringes of Fig. 6.50 are not straight because the sensitivity to displacement varies along the bar. A high k-value represents a low sensitivity. The calculated displacement at D must, of course, be the same if the fringes are counted along CD, along CEFD, or along CEGD.

displacement of D normal to the ellipsoids of the holo-diagram can be found by multiplying this number by the k-value (11) at D. Care must be taken not to count the same fringe twice with the same sign. The number of fringes should always be independent of the chosen path. For example, the fringes along the path CEGD should be counted in the following way: CE = 0, EG = +9, GD = −3. The sum, six fringes, is identical to the result obtained by tracing CD directly. This rule of sign is obvious when the continuous fringes of Fig. 6.50 are studied, but the sign of the fringes might be difficult to find if only fractions of the fringes are visible, e.g. only those along the edges EGD of Fig. 6.51.

If one makes the mistake of multiplying by the k-value at the point where a fringe is crossed, the value of the displacement will depend on the chosen path. For example, follow the path CEFD, which produces the following counts: CE = 0; EF = 6; FD (along one fringe) = 0. The counted fringes (six) should be multiplied by the k-value (11) at D, *not* by the k-value (7) at EF where all the fringes were crossed.

6.6.4 Compensating for large in-plane motions by sandwich holography

Before finishing this chapter let us finally consider how to evaluate a small deformation of an object that has made a large translation.[7] Just as sandwich holography can be used to eliminate the fringe pattern caused by large object tilts (see Figs 6.19 and 6.20 for example), it can also be used to compensate

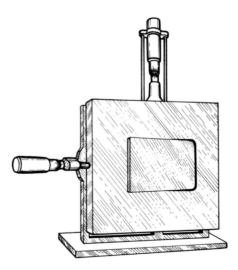

Fig. 6.52. The test object, a steel plate that can be translated in the plane. In the centre is a door that can be tilted around a vertical axis fixed to the plate. This means that the door can be both translated and tilted at the same time.

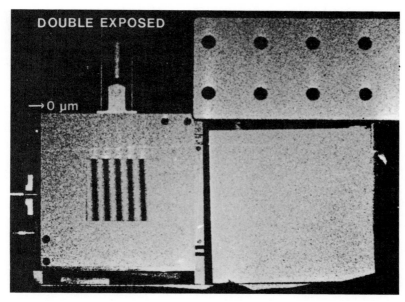

DOUBLE EXPOSED

→ 0 µm

Fig. 6.53. Double-exposure hologram of the object in Fig. 6.52. A tilt of the door between exposures causes the vertical fringes seen on the rectangular, inner part of the object.

for large in-plane motions. Thus it is also possible in this case to separate a small local deformation from a large rigid body motion. In Section 5.4 the equation (equation (5.16)) was derived for the fringes formed by in-plane motions (or more strictly speaking by motions perpendicular to the line of sight). The sandwich tilt needed to eliminate the effects of an unwanted rigid body motion of the object is given by equation (5.37). In the following a practical experiment will be presented which verifies these statements.

The test object consists of a steel plate that could be translated in its plane (Fig. 6.52). At its centre there is a door which can be rotated about a vertical axis fixed to the plate. The design is such that the door, in a controlled way, can be both translated with the steel plate and tilted in relation to it at the same time. Figure 6.53 shows the result of a doubly exposed hologram when there was no in-plane motion but only a tilt of the door in between the two exposures. Five vertical fringes have formed on the rectangular inner part of the test object which represents the door.

In Fig. 6.54 the plate with the door has been translated horizontally not less than 1 mm while the door underwent the same tilt as in Fig. 6.53. No fringes are seen on the frame of the door because the large translation produced such a high number of fringes (about 250) that they are not resolved. Figure 6.54, however, is a photograph of a reconstruction from a

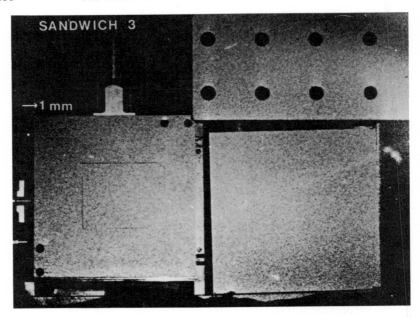

Fig. 6.54. Sandwich hologram. A translation of the whole object by no less than 1 mm combined with a tilt of the door. The reference surface is fringe-free; the object is covered by fringes that are so closely spaced that they cannot be seen.

sandwich hologram: by tilting this hologram it was possible to eliminate the fringes caused by the 1 mm translation (see Fig. 6.55).

Before the sandwich hologram was tilted no fringes could be seen anywhere (Fig. 6.54), but as soon as the tilt was introduced fringes appeared on the fixed reference surface (the right-hand square). As the hologram tilt was increased the fringe frequency grew higher until the fringes became so closely spaced that they could be resolved no longer. At still larger tilts fringes of high frequency began to appear at the translated steel plate with its door. These fringes moved as the observer's eye was moved behind the plate.

The sandwich tilt was so adjusted that the fringe frequency decreased and finally became zero, so that no fringe, or just one fringe, was seen on the frame. Under these conditions the fringe parallax disappeared too. Thus the fringes on the door became stationary at the same time as the frame became fringe-free, and also remained fringe-free as the eye was moved behind the plate.

The fringes seen on the door (Fig. 6.55) were identical to those seen in the double exposure of Fig. 6.53 in spite of the fact that a translation of 1 mm had been compensated for. Thus it has been verified that a local out-of-plane displacement of c. 1.5 µm can be isolated from the large in-plane motion of

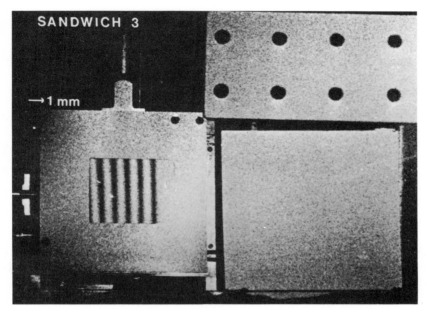

Fig. 6.55. The same sandwich hologram as in Fig. 6.54, now tilted in such a way that the 1 mm translation has been compensated for. The fringes at the door reveal a tilt of less than 2 μm in spite of the rigid translation of 1 mm.

1000 μm. That the fringes caused by the latter motion had been eliminated was indicated by the disappearance of the fringe parallax.

References

1. N. Abramson. Sandwich hologram interferometry. 2. Some practical calculations. *Appl. Opt.* **14**, 981 (1975).
2. W. Schumann and M. Dubas, *Holographic Interferometry*. Springer Verlag, New York (1979), p. 69.
3. N. Abramson. In *Proceedings: Electro-optics International Conference Brighton 1974*. Kiever Communications Ltd, Surrey, England (1974), p. 35.
4. N. Abramson. Sandwich hologram interferometry. 4. Holographic studies of two milling machines. *Appl. Opt.* **18**, 2521 (1977).
5. H. Bjelkhagen. Pulsed sandwich holography. *Appl. Opt.* **16**, 1727 (1977).
6. N. Abramson and H. Bjelkhagen. Pulsed sandwich holography. 2. Practical applications. *Appl. Opt.* **17**, 187 (1978).
7. N. Abramson and H. Bjelkhagen. Sandwich hologram interferometry. 5. Measurement of in-plane displacement and compensation for rigid body motion. *Appl. Opt.* **18**, 2870 (1979).

7

Equipment and methods

Optical holography is based on interference between light beams during exposure and diffraction during the reconstruction process. Interference always occurs when two light beams intersect, but in order to produce stationary interference fringes the two beams have to be coherent, to oscillate in a well ordered way.

Using the moiré pattern analogy of Fig. 1.5, the wavefronts are represented by lines that are perpendicular to the two intersecting beams AD and BC respectively. An elementary condition for the moiré fringes to be stationary when the wavefronts move along the light beams is that they are straight and equally spaced in the two beams. In that case each beam is said to have ideal transverse (spatial) and longitudinal (temporal) coherence. The two identical beams are also said to be mutually coherent.

The only practical way to produce two mutually coherent beams for holographic work is by using one laser, the single beam of which is split into two beams: the object beam, and the reference beam which is necessary to make the hologram plate phase-sensitive. The splitting of the laser beam can either be carried out by using different areas of the expanded beam (single beam holography with wavefront division) or by reflection from a semi-transparent mirror (amplitude division). Whenever possible the former method is used because of its simplicity and lower sensitivity to disturbances (see Section 3.3.1).

7.1 Lasers

For holographic work the laser has enormous advantages by comparison with any other light source because it produces a beam of light with a high degree of spatial and temporal coherence. When Dennis Gabor proposed the theory of holography in 1947 he was hindered in providing practical proof by the fact that the laser had not been invented. He had to use a mercury lamp instead and to increase the transverse coherence by spatial filtering while the temporal coherence was increased by interference filters that only let through a single spectral line. These devices cut out most of the light and therefore the resulting light intensity became so low that only small objects could be studied. The temporal coherence was in any case severely limited, which resulted in the fact that the depth of field was severely restricted too.

Coherence length is a measure of the temporal coherence. If there is a minute change in the separation of the wavefronts in the beam AD in Fig. 1.5, the number of waves per unit length will still be the same over small distances. Over larger distances, however, the small variations can accumulate to such a degree that there could be half a wave too much or too little compared to the perfect beam BC. In that case the phase error would be 180° and the moiré fringe will have moved so that the dark fringe is replaced by a bright one. If these phase shifts are random, the moiré fringes will move around when the wavefronts travel to the right, so that they become blurred out. That will also be the case with the corresponding interference fringe. If the laser beam is split in two and if the path lengths differ by more than the coherence length, then no fringes will be seen when the two beams are reunited.

A simple demonstration of the importance of coherence length is given in Fig. 7.1. A laser beam is divided by the beam-splitter (B) so that one component is reflected directly towards D, while the other component passes through B and is reflected by the mirror C before it also reaches D. If fringes are formed at D the distance BC is shorter than the coherence length of the laser light. This length varies for different types of lasers from a few

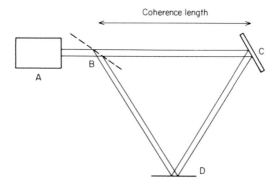

Fig. 7.1. One part of the beam from a laser (A) is reflected by the beam-splitter (B) to the screen (D). The other part of the beam passes through (B) to the mirror (C) which reflects it to D. If the distance BC is shorter than the coherence length, interference fringes are formed at D.

centimetres for the simplest ruby lasers to some 30 cm for the simple He–Ne laser and to tens of metres for the best argon lasers.

7.1.1 He–Ne laser

Let us start by considering the He–Ne laser because it is by far the most common laser currently used in holographic work. LASER stands for Light Amplification by Stimulated Emission of Radiation. It is characteristic of the light emitted by any laser that it can be intensive, parallel, monochromatic and coherent—all at the same time.

Let us scrutinize more closely each of the letters in the word LASER. The laser effect is by no means limited to the electromagnetic radiation that can be seen by the eye. The first laser was not even a laser at all but a MASER, where the M stands for Microwave. There now exist lasers with outputs which range from the long wavelengths of radio waves through infrared light, across the whole visual spectrum and down to the short waves of ultraviolet light. Indeed there are good reasons to believe that in the near future X-ray lasers and gamma-ray lasers will become available.

The laser is an optical device which amplifies light. There is of course no problem in deamplifying light: even the cheapest pair of sunglasses reduces the intensity of light passing through them. To amplify light, however, the phenomenon of fluorescence has to be invoked. Some materials have the property of giving off light a short while after they have absorbed light energy. Such materials include the zinc sulphide used to paint the hands and figures of some watches, the rare gas neon (Ne) used in neon signs, and the semi-precious gems called rubies which consist of alumina (Al_2O_3) together with small traces of chromium ions which produce the red colour.

When light passes through neon it is absorbed at certain wavelengths (certain spectral lines) at which the frequency of the light is in resonance with the atomic vibrations. These wavelengths represent particular energy levels in the atom because there exists a simple relation

$$e = h\nu, \qquad (7.1)$$

where e is the energy of one photon while h is Planck's constant and ν is the frequency of the light.

When neon is heated, e.g. by an electric discharge, it starts emitting light at exactly those wavelengths (spectral lines) at which it absorbs light when it is at room temperature. The energy needed to produce light is provided by the electric current; this energy sets the atoms in violent motion and their electrons consequently move into more energetic orbits. This process is called "pumping" of the laser system because energy is pumped into the laser material, the atoms of which are brought into an excited state.

After a while the electrons spontaneously drop down again into a lower energy orbit, giving off a photon with the particular resonance frequency that corresponds to the energy difference between the two orbits. There is a small delay between the absorption of energy by an atom and its subsequent emission of a photon; during this period the atom is willing to give off its light energy but does not know when or how. It is said that the atom belongs to an inverted population because it is in an unnatural, metastable, and energy-rich state.

If the atoms of an inverted population are hit by light of the same wavelength as they were going to emit spontaneously, the atoms are immediately triggered to emit their own energy. In this way another, identical photon has been added to the triggering photon: the impinging light has been amplified. This effect is known as *stimulated* emission to differentiate it from the spontaneous emission already described. The existence of this pheno-menon was predicted theoretically by Albert Einstein when he calculated the energy equilibrium between gas atoms and light quanta in a closed reflecting box. The probability of absorption and the probability of emission should be equal, but as spontaneous emission is delayed in such a box (and absorption is not) there had to exist an effect that speeded it up. This effect was the stimulated emission.

Thus a pumped laser medium will amplify light in a sort of chain reaction, similar in some ways to that of an atomic reactor (a continuous laser) or an atomic bomb (a pulsed laser). The amplifying mechanism is non-linear, so that strong light is amplified more than weak light. The amplification of course ceases to exist when the population is no longer inverted. However, as long as there is an inverted population the emission can be larger than the absorption and therefore the more gas which is looked through the brighter it will appear. A long straight tube designed like a neon sign would give out quite a strong light at its ends. (There exist lasers, e.g. the pulsed nitrogen laser, that are built in this simple way. The pulsed nitrogen laser produces its ultraviolet output beam by means of the effect known as super-radiation.)

If mirrors are installed at the ends of a tube filled with an inverted population of neon atoms they will produce a similar effect to that of the tube being much longer. If the two mirrors are also adjusted with such high precision that they are exactly parallel to each other and perpendicular to the tube axis, then the photons can be reflected back and forth between the mirrors. The light would then be amplified until the losses are equal to the amplification. The neon sign has become transformed into a laser and the space between the mirrors is known as a *cavity*.

The introduction of the mirrors results in radical changes in the properties of the light inside the cavity. Because of the feedback effect of the mirrors, the amplifier, which previously only amplified noise, is transformed into an oscillator of high quality. Only those photons that initially move parallel to the tube axis can be reflected back and forth by the mirrors, and thus only the light in the axial direction will be amplified. As the amplification is a non-linear process these light rays will "win" over the rays in other directions and take most of the energy from the atoms.

As the light is reflected back and forth along the tube, standing waves will form inside the cavity. As one particular spectral line will be best suited for the laser action it will have enhanced energy with respect to the others. The result will be that strong monochromatic light is radiated along the cavity axis. Finally, a laser beam will pass out through one of the mirrors, which is semi-transparent.

In an ideal situation the laser beam would consist of light having a single, infinitely thin spectral line. However, because of the effect of temperature, the atoms in the gas travel in all directions at velocities measured in kilometres per second, thus there is a doppler broadening of the spectral line. Although the atoms oscillate with very constant frequency, the wavelength of the light will vary. The doppler effect means that an atom moving towards the observer will produce light of a shorter wavelength and vice versa. For this reason the laser will emit light having a narrow band of wavelengths. However, even if this variation in wavelength is extremely small the effect shortens the coherence length of a He–Ne laser to about 30 cm and that of an argon laser to about 5 cm (the gas of the argon laser is hotter).

7.1.2 Coherence length

The wavelength of light from a laser is not only determined by the spectral lines of the atoms but also by the cavity length. The reason is that only those light waves are amplified that arrive in phase when they coincide with themselves after having been reflected once by each mirror. All waves that arrive out of phase will be eliminated by destructive interference (Section 1.1.2 and Fig. 7.2). Another way to say the same thing is that twice the cavity length ($2L$) should be an integral multiple (n) of the wavelength (λ). Thus the following equation is obtained:

$$n\lambda = 2L. \tag{7.2}$$

Replacing λ by c/f, where c is the speed of light and f the frequency, and differentiating leads to the following result:

$$df = \frac{c}{2L}\,dn. \tag{7.3}$$

Thus there exist many frequencies ($dn = 1, 2, 3$, etc.) with the difference df. The light intensity will therefore vary with this difference frequency, i.e. there will be a beating effect in the output. The "wavelength" of these "beat waves" will be equal to twice the cavity length ($2L$).

The two restrictions on laser output can be summarized in the analogy

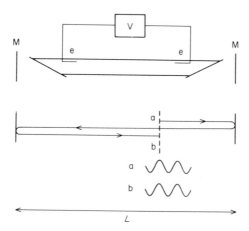

Fig. 7.2. A gas laser consists of a tube containing gas, two electrodes (e) connected to a high voltage supply (V) and two mirrors (M) separated by the distance L. A light wave (a) passing a certain point is reflected first by the right-hand mirror and then by the left-hand mirror, after which it arrives back at the same point again (b). It is necessary for the laser to function that the phase is identical at (a) and (b).

illustrated in Fig. 7.3. Only those frequencies will appear in the laser light that are within the doppler-broadened spectral line and at the same time fulfil the conditions set by equation (7.2). As the spectral width is independent of the cavity length (L) and df is inversely proportional to L, the cavity length can be introduced into Fig. 7.3 assuming instead that df is independent of L while the spectral width is a direct function of L.

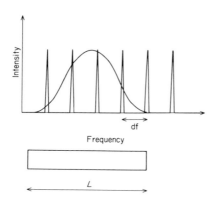

Fig. 7.3. To fulfil the conditions of Fig. 7.2 only certain wavelengths, and thus certain frequencies, are emitted by the laser. The sharp peaks separated by df represent these frequencies while the smooth curve represents the doppler width of the spectral line. The longer the laser cavity (L) the more frequencies are within the doppler curve.

Using this analogy it can be seen that a laser of large cavity length will produce many frequencies while a short laser ($L \sim 10 \, \text{cm}$) will produce just one frequency (a single-frequency laser). The latter can have an almost infinite coherence length (many kilometres) whereas the longer lasers have a coherence length determined by the spectral line. The short, single-frequency laser is not suitable for practical holography because of its low output (less than $1 \, \text{mW}$). The reason for this low output is that while the losses at mirrors and windows are constant the gain is proportional to the length of the gas column used.

To transform an ordinary, long, multi-frequency laser into a single-frequency laser one more wavelength-selective element is needed in addition to the atomic resonances and the cavity length. The most common method employed is that of introducing a further small cavity so that the laser light is limited to those frequencies that are integral multiples of the wavelengths in both the large and the small cavity.

Such a short extra cavity is strictly referred to as a Fabry–Perot étalon, but is usually known more simply as an étalon. It consists of one thick parallel flat plate of glass or of two parallel glass plates separated by air. The effect of its inclusion is seen in Fig. 7.4. In order not to introduce more than one extra cavity the étalon is usually set at an angle so that there will be no multiple

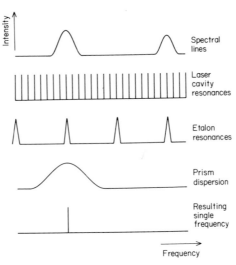

Fig. 7.4. The argon gas emits light only at some spectral lines. The length of the laser cavity restricts the emitted light to one set of frequencies. The thickness of the étalon restricts the light to another set of more widely spaced frequencies. Finally, the dispersion of a prism, or the diffraction of a gitter, selects one of the spectral lines. Only a single frequency will pass through all these wavelength-selective components to produce the laser beam.

reflections between the laser mirrors and the étalon surfaces. If the distance between the two laser mirrors is changed, e.g. increased, the resonance frequencies will move to the left. An increased tilt of the étalon will result in a movement of the selected wavelength to the left. In order to maintain constant coincidence between the two resonant frequencies within the doppler curve either the mirror separation must be kept constant with a high degree of accuracy (e.g. temperature control) or the frequencies have to be adjusted by positive feedback so that the output does not vary (i.e. does not make any mode jumps).

7.1.3 Argon and ruby lasers

In addition to the three wavelength-selecting elements already discussed, the argon laser needs a fourth element in order to be able to emit true single-frequency light. The reason is that the rare gas argon provides more than one spectral line that has the right gain and the long lifetime of the excited atoms needed to be able to "lase". Thus an argon laser, in spite of being equipped with an étalon, can still emit light of more than one wavelength. The remedy is to introduce a prism or a grating in the cavity, as shown in Fig. 7.5.

Because of its dispersion effect, a prism refracts light of different wavelengths at different angles, therefore the mirror M_2 appears to be perpendicular to the cavity only for one certain colour and only waves of corresponding length can be reflected back and forth between M_1 and M_2. Thus only light of a certain colour will "lase" for each setting of the prism (P) or the mirror (M_2). Often there is no independent mirror but the back of the prism is coated to make it reflective.

The prism could be replaced by a grating and this grating could even replace the mirror M_2, so that one end of the cavity could be constituted by the grating itself. The prism solution is, however, most common as it is simple, durable and causes only small losses.

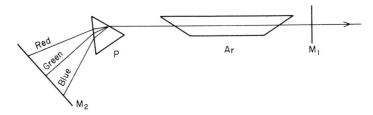

Fig. 7.5. The argon laser has a prism (P) inserted between the two mirrors (M_1 and M_2) of the cavity. Because different colours are refracted at different angles the mirrors will appear optically parallel to only a single colour. By tilting one mirror, for example, the wavelength of the laser light can be monitored.

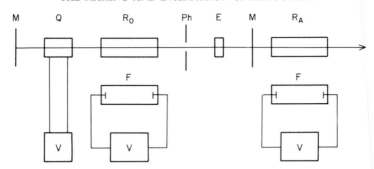

Fig. 7.6. Example of a ruby laser designed for holography, R_O and R_A are the oscillator ruby and the amplifier ruby respectively. F represents the flashlamps with their high voltage supplies (V). M–M represents the mirrors forming the cavity of the oscillator ruby with its pocket cell (Q), pinhole (Ph) and étalon (E).

The design of a ruby laser (Fig. 7.6) differs from that of a gas laser because it cannot be pumped by a discharge in the non-conducting ruby. Instead it is pumped by the light from a gas discharge lamp similar to the electronic flashlights used in ordinary photography. The pump light is absorbed by the chromium atoms which thus become excited and for a short while thereafter are able to amplify the fluorescent light which they would otherwise emit spontaneously. A very high pumping intensity is needed and therefore almost all ruby lasers work in short pulses (in contrast to the continuous wave gas lasers). The design of the cavity with its frequency-determining elements is basically similar to that of gas lasers. However, a special shutter (Q) is needed in order to produce the desired pulse-forms suited for holography.

The firing of a ruby laser for holography usually proceeds in the following way. First, large capacitors have to be charged, which takes some 30 s (Fig. 7.6). Then the gas discharge is triggered either manually or by a signal from the event to be holographed. This lamp will emit a light pulse of a few milliseconds in duration. During this time the ruby is pumped and, if there is no shutter, it will start producing laser pulses as soon as the gain is larger than the losses of the cavity. Usually a train of some hundred or so pulses is produced as the result of a single lamp flash. However, these free-running pulses are not suitable for holography because of their random distribution and short coherence length.

When the shutter is closed there will of course be very high losses in the cavity as the light is prevented from reaching one of the mirrors and therefore no laser pulse is formed. Thus the pumping can go on for a longer time without the atoms giving up their energy and therefore more light energy is accumulated. If the shutter is then suddenly opened for a short period of time, one single pulse of extremely high energy is produced. The intensity can be hundreds of times higher than observed without the shutter. This type of laser

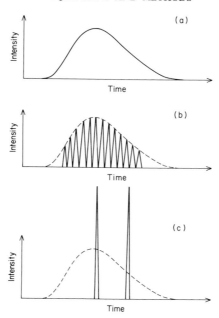

Fig. 7.7. (a) The intensity of the pumping flashlamp as a function of time; (b) the multiple pulses from a conventional ruby laser; (c) the two giant pulses from a Q-switched laser for double-exposure holography.

pulse is called a "giant pulse" and the method described is known as "Q-switching". The letter Q refers to a gain factor representing the relationship between amplification and losses. If the shutter is opened twice during one flash from the discharge lamp, then two laser pulses are produced which are suitable for making a double-exposure hologram (Fig. 7.7).

The electronic shutter can take the form of a Kerr cell (using a liquid as the active medium) or a Pockels cell (using a crystal). In both cases the polarization of the laser pulse is rotated by electrostatic fields so that it can pass through two crossed polaroid filters that otherwise would stop it. The Q-switched pulse usually has a duration of some 20 ns (20×10^{-9} s) and the time separation between the pulses can usually be varied from 300 to 700 μs.

If only one pulse is needed the expensive electronic Q-switch can be replaced by a passive Q-switch which consists simply of a cuvette filled with a bleachable dye, e.g. rhodamine. This semi-transparent dye stops the laser from "lasing" until the population is so heavily inverted that the amplification becomes larger than the losses despite the presence of the dye cell. As soon as the laser action starts the dye is bleached, and thus the losses suddenly decrease to a low value. The action of the dye cell is not only that of

Q-switching the laser, it also usually substantially improves the coherence length of the light as compared to the result using a Pockels cell.

The commonest wavelength for ruby lasers is 694.3 nm while the strongest lines of an argon laser are 488.0 nm and 514.5 nm.

7.1.4 The influence of polarization

There exist two types of He–Ne laser. In one of these the two ends of the tube are sealed by inclined windows outside which the cavity mirrors are mounted. In the other type the mirrors are fixed within the partial vacuum, directly on the inside surface of the windows (internal mirrors). The latter type has many advantages as it results in a very robust and endurable laser that can be used in vibrating and dusty environments.

The lasers with separate windows usually have their mirrors mounted so that they can be aligned by turning adjustment screws. The space between the windows and the mirrors is usually sealed by metal bellows to prevent contamination by dust particles. The gain of a He–Ne laser is usually only a few per cent and therefore it is important to keep the losses at windows and mirrors extremely low.

An ordinary glass window results in losses of the order of 5% due to normal reflection, and this is much too high to be acceptable. Therefore the ends of the tube are cut at an angle of around 34° and thus the windows are mounted so that their normal is inclined by 56° to the cavity axis. At this very special angle (the Brewster angle) the reflection losses for a particular polarization of the light are close to zero while the losses in the perpendicular polarization angle remain high. Thus the gain-to-loss ratio is high for only one polarization angle and the laser light will be almost completely linearly polarized.

The holographic set-up is usually arranged on a horizontal surface so that vertical deflections of the laser light are much less common than horizontal ones. For this reason most lasers used in holographic work are so designed that the normal to the Brewster windows are in a vertical plane. In this way there is no risk that a mirror deflecting the light horizontally (e.g. to produce the reference beam) could accidentally be positioned at the Brewster angle. At this angle the reflection from a clean glass surface is cut off completely while reflections from beam-splitters and ordinary mirrors are less influenced.

In this connection it might be important to point out that only light beams polarized at identical angles can produce high contrast interference fringes. If, therefore, the object beam is polarized vertically (i.e. the electrical field component oscillates vertically) while the reference beam is polarized horizontally, no hologram will be recorded.

In some rare cases the object, or some optical element, can rotate the polarization, resulting in low quality holograms. One way to look for this possible error is to see whether the object beam and the reference beam are cut out at the same angle when they are studied through a polarizing filter which is slowly rotated. If this is not the case, then one beam could be rotated by a rotator (a quarter-wave plate). If the light that is scattered by the object becomes randomly polarized, it might be advantageous to place a polarizing filter in front of the hologram plate so that light of incorrect polarization does not darken the plate unnecessarily.

The other type of laser, having internal mirrors and no Brewster windows, is usually said to produce unpolarized light. This statement is, however, not quite true. Instead, every second mode of the axial modes under the doppler curve (Fig. 7.3) is polarized in one direction while the adjacent modes in between are polarized in the perpendicular direction. When the laser heats up and expands, the axial modes will move to the left, and therefore one polarization state will alternate with the other as they pass the peak of the doppler gain curve. If the laser is long the variation in polarization will not be pronounced, but if it is so short (e.g. 30 cm) that only some three or four modes exist within the doppler curve, then the variation between the two polarization states will be almost total.

When such a laser beam is divided into two by a beam-splitter the intensity of the reflected light can vary considerably if the semi-transparent mirror is close to the Brewster angle. The present author does not recommend use of a gas laser without Brewster windows for holographic work, because the ratio of the intensities of the reference and object beams will vary in a random manner. It is advantageous to combine the good protection of the sealed mirrors with a fixed polarization. The two types of laser are therefore often combined so that a laser with internal mirrors is also supplied with a Brewster window, the sole function of which is to produce a fixed direction of polarization.

Usually the direction of polarization is of no importance as long as it is the same for the object and the reference beam when they arrive at the hologram plate. Sometimes, however, the quality of the holographic recording is lowered because the plate becomes covered by broad interference fringes caused by the reference beam being internally reflected back and forth between the surfaces of the glass plate. One way to eliminate this process is to cover the rear side of the plate with an absorbing paint, usually referred to as an "antihalo coating". The present author has used ordinary black shoe-polish for this purpose with good results. This method is, however, rather messy and has the additional limitation that it cannot be used for "Lippmann–Denisyuk holograms" at all because in such holograms light must enter from both sides of the plate.

One solution to this problem is to position the hologram plate so that the reference beam enters at the Brewster angle. The best way to accomplish this is by adjusting the angle of the plate until it does not reflect the reference beam—or at least until the reflection is at minimum. This method works well, even if it does sometimes require a quarter-wave plate to be introduced into the beam or the laser to be laid down on its side.

7.1.5 Practical considerations

The gas of a He–Ne laser is a mixture of 86% helium, used for the energy transfer, and 14% neon, which produces the laser light. The gas pressure should be a few millimetres of mercury. As the laser is used, gas is consumed at the electrodes and therefore the tube of the laser should be so designed that it holds a large volume of gas. Usually the laser is delivered with quite a high gas pressure and its lifetime is over when the pressure has decreased too far. When the laser is not being used atmospheric gases can leak in, leading to an increase in the pressure.

Thus a laser should not be left to stand on the shelf for months. It should be used now and then, e.g. switched on once a week, in order to be kept in good condition. If a laser does not produce a beam in spite of the fact that its tube is glowing, or if the beam is very weak, there are good reasons to leave it on for some hours to see if it will recover. If the red colour of the gas discharge has got an unusually strong blue component, then the tube probably has to be exchanged. A good laser should nowadays have a lifetime of some 10 000 h.

The discharge in the laser tube is initiated by a pulse of some 20 000 V while the running voltage is only around 1000 V. The starting device is similar to that used in ordinary neon signs or fluorescent tubes and the starting cycle is usually totally automated.

The mirrors of the laser have to be of high efficiency. The losses of an ordinary metal mirror are of the order of 5%, which is much too high compared to the gain of the laser, which is only a few per cent ($5-15\% \, m^{-1}$). Therefore the reflecting metal film of ordinary mirrors cannot be used, instead the glass plate is coated by a number of thin films of materials having alternating high and low refractive indices (e.g. magnesium fluoride and titanium dioxide respectively). Using some 20 layers the reflectance can be of the order of 99.85%, which is quite a sufficiently high value as compared to the gain of the He–Ne laser.

When the laser cavity is composed of two flat parallel mirrors, the accuracy of the relative positions of the two mirrors which is required is such that it is difficult to handle the laser tube at all!

The sensitivity to small tilts of one mirror is much lower if, instead, the two mirrors are spherically concave with the centre of curvature of each mirror almost coinciding with the surface of the opposite mirror. A light ray can then still be reflected back and forth between the mirrors even if the path is slightly altered on each reflection. The disadvantage of such a mirror configuration (confocal cavity) is that there is a chance that a ray can take more than one path and thus that more than one transverse mode is formed.

A configuration in which one mirror is flat while the other is spherically concave combines a good spatial coherence (only one transverse mode) with a low sensitivity to errors in the mirror alignment. The separation of the mirrors should be slightly smaller than the radius of curvature of the concave mirror. This type of cavity is referred to as hemispherical. There also exist other combinations of mirrors, e.g. two identical concave mirrors separated by a distance that is slightly larger than twice the radius of curvature. Because of their larger separation compared to the confocal configuration described above, the spatial coherence is better.

Lasers are required to produce parallel beams of light and therefore those lasers that have spherical mirrors need compensating elements to make the divergent rays parallel. The reflective coating is on the inside of the cavity mirrors, so the correction of the rays can be achieved by making the outside surface of the mirror in the form of a lens surface.

Even if the outside surface of the mirror is antireflection coated, it still produces a weak beam that is reflected back against the mirror surface and again reflected out of the laser. Therefore the outside surface of the output mirror is tilted so that this unwanted extra laser beam is reflected at an angle to the main beam, producing a weak extra laser spot.

Most modern lasers do not need any adjustments or cleaning of their mirrors. If they do, the instructions given by the manufacturer should be followed with care. It is extremely easy to destroy the delicate coatings of a laser mirror by using an unsuitable cleaning method.

7.2 Beam expander

When the thin laser beam is to be used for holography it has to be "expanded" enough to illuminate the whole object. This expansion of the beam can be accomplished by either a negative or a positive lens. Usually one wants the beam to diverge rather quickly. A divergence of 0.2 rad might be suitable, which means that the beam diameter becomes close to 0.2 m at a distance of 1 m from the laser cavity.

Let us assume the original beam to have a diameter of 1 mm and a divergence of 0.5 mrad, meaning that the diameter increases to 0.5 m at a distance of 1000 m. However, after passing a lens the diameter of the beam is required to be 0.2 m at a distance of 1 m (Fig. 7.8). The focal length of the lens can be found using simple trigonometry:

$$f = d_1 l/d_3, \tag{7.4}$$

where f is the focal length, d_1 the diameter of the original laser beam, d_3 the

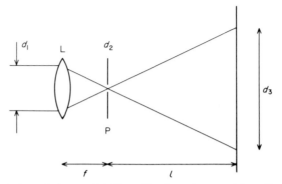

Fig. 7.8. Calculation of the spatial filter. The laser beam with a diameter d_1 enters from the left. It is focused by the lens (L), with focal length f, to the pinhole (P) with the diameter d_2. Thereafter it diverges to the diameter d_3 at the distance l.

diameter of the expanded beam, and l the distance from the focal point of the lens to the object. Thus in the present case the focal length of the lens should be 5 mm. Usually a microscope objective is used instead of the single lens of Fig. 7.8. The magnification power of this lens is expressed as 18 cm (normal viewing distance) divided by f, resulting in a power of $36 \times$ for the particular objective in question.

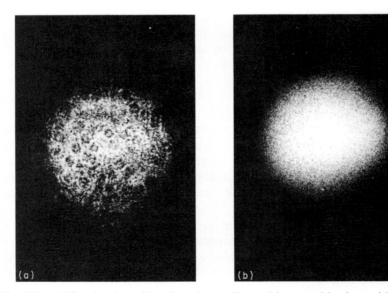

Fig. 7.9. (a) Illumination with a laser beam diverged by a positive lens without the use of a spatial filter: the spatial noise in the form of spots, lines and circles is caused by scratches and dust at the optical surfaces. (b) The same illumination after spatial filtering by passing through a pinhole placed at the focus of the lens.

7.2.1 Spatial filter

The quality of the expanded beam will be much improved if an aperture with a pinhole (P in Fig. 7.8) is positioned at the focal point of the lens. When there are scratches and dust at the lens surface or on the optical parts of the laser, the beam will detoriate as a consequence of diffracted and scattered light rays which are not travelling in directions parallel to the beam. These stray rays will produce both constructive and destructive interference, causing the main beam to be badly affected by spatial noise consisting of dark and bright spots and fringes (Fig. 7.9).

However, if a sufficiently small pinhole is placed at the focal point of the lens, all the light that is travelling in the wrong direction will be cut off and only a clean uniform laser beam will pass through. To gain this result the size of the pinhole has to be so small that even rays diffracted by its edges cannot cause destructive interference within the part of the expanded beam which is to be used. This condition guarantees that no spatial noise can pass through the pinhole area. Thus the diameter d_2 of the pinhole (see Fig. 7.8) should be such that no Young's fringe can form within the illuminated area d_3. From equation (1.7) it is found that

$$d_3 = \frac{\lambda}{2 \sin \alpha} \simeq \frac{\lambda l}{d_2} ;$$

combining this result with equation (7.4) produces the relationship

$$d_2 = \lambda f / d_1 . \tag{7.5}$$

This value, when multiplied by 1.22, is equal to that given by the definition of the diffraction-limited spot size of a lens. It is common to use double this value for the pinhole diameter, in order to arrive at a practical balance between losses and noise suppression.

Thus the pinhole size should satisfy the equation

$$d_2 = 2.44 \lambda f / d_1 \tag{7.6}$$

where d_2 is the pinhole diameter, λ the wavelength of laser light, f the focal length of the lens (180 divided by magnifying power), and d_1 the diameter of the original laser beam.

A laser beam has no sharp edges and it has a Gaussian intensity distribution. The radius of the beam is defined as the distance from the beam axis to the point at which the intensity is $1/e^2 = 0.135$ times the central intensity. The diameter d_1 of equation (7.6) is equal to twice this value and that same definition of beam diameter is used in laser data sheets.

A spatial filter is important for high quality holography and its position in the holographic set-up is demonstrated in Figs 3.14 and 3.15. The filter must be mounted with high stability because the beam must enter the centre of the pinhole and also because the pinhole represents point A of the holo-diagram and thus a point that should be consistently determined with interferometric precision.

Aligning the spatial filter to the beam can be difficult the first time it is attempted, and the following procedure is recommended. Mark the position of the original laser beam on a fixed paper screen. Then insert the spatial filter with its pinhole removed and adjust it to be centred in the beam. The expanded laser beam should then be rotationally symmetric and centred around the original beam. The reflection from the lens should return close to the output mirror of the laser. Thereafter the pinhole is inserted and a sheet of paper placed close to the spatial filter. Look for a small red spot while translating the pinhole across the beam by using the adjustment screws. It is, of course, easier to find the position at which the beam enters the pinhole if the latter is placed at a point where the beam is wider than it is at the focal point. Therefore the pinhole should first be placed close to the lens until the laser spot is found. It should then be moved towards the focus, which is found as the position at which most light passes through the pinhole. The closer to the focal point the more important is the alignment, therefore repeated adjustments have to be made. When the adjustment is correct the illumination field should appear uniform without any spatial disturbances. If the pinhole is too large there will remain some large scale variations in light intensity. If too small, much of the intensity will disappear. The losses caused by passing the beam through the pinhole should be less than 10%.

Generally, it is not possible to use a spatial filter when a pulsed ruby laser is used as a light source. The power concentration becomes so high at the focus of the lens that the air molecules are broken down. The plasma cloud formed in this way is opaque to light and thus stops its further propagation. The spatial filter could, of course, contain a more suitable gas or even a partial vacuum, but no device based on such a solution is commercially available at the time of writing and therefore the beam has to be expanded by a diverging lens instead of a converging one. Thus no pinhole can be used and spatial noise cannot be eliminated. This is one of the main reasons why holograms made with the use of ruby lasers are usually of relatively low quality. The only remedy is to keep the laser optics and the diverging lens as clean as possible.

7.3 Utilizing the coherence length

The coherence length has already been defined as the maximum path length difference between two laser beams from the same laser that still results in acceptable interference fringes when they are recombined (Fig. 7.1).

The main reason for the limitation of the coherence length is the doppler broadening of the spectral lines. There exist many ingenious methods for increasing the coherence length, as shown in Fig. 7.4. Often, however, one has to work with a conventional He–Ne laser and to make the best of the light available. By carefully planning the holographic configuration it is possible to record objects that are more than 1 m long even if the coherence length is only some 15 cm.

To optimize the utilization of the coherence length in a holographic set-up the path lengths of the object beam and the reference beam should be carefully measured and made equal. If no beam-splitter is used, as in the configuration of Fig. 3.15, a string can be stretched from the spatial filter to the centre of the reference mirror and from there to the centre of the hologram plate, where the string should be marked with a knot. Thereafter the same procedure can be repeated, but this time the string should pass via the centre of the object instead of the reference mirror. The difference in string length between the two measurements represents the difference in path length and should be no greater than the coherence length. If the object is large the procedure has to be carried out for the points on the object which are farthest from and closest to the source of illumination and the point of observation in order to get it all within the coherence length.

Another method which can be used for this comparison is as follows. First, mark off on the string the path length via the reference mirror. Thereafter, shorten the string by one coherence length and confirm that no part of the object can be reached by the string when its ends are fixed to the spatial filter and the hologram plate respectively. Then lengthen the string, instead, by one coherence length and see that every point on the object can be reached by the string. In this way it is assured that the entire object is within the coherence length and thus can be recorded. The next section will show that the described method is identical to using the ellipsoids of the holo-diagram.

7.3.1 Using the holo-diagram

If the two ends of a string are fixed by nails to a flat piece of paper and a pencil is moved along the string, keeping it taut all the time, the pencil will draw an ellipse. Thus the procedure for measuring path lengths described in the preceding section is identical to placing the object between two ellipses having their focal points at the point of illumination (the spatial filter, A) and the point of observation (the hologram plate, B). This procedure leads back to the use of the holo-diagram[1] first introduced in Section 1.4.1.

Let us study Fig. 7.10. The reference beam travels from A to B via the mirror M, which reflects the light in the required direction only if M is tangential to the ellipse. If a small object (C) is placed on the same ellipse as the mirror, the need for a large coherence length is almost zero. A long toy train curved along that particular ellipse could be recorded even if it was a

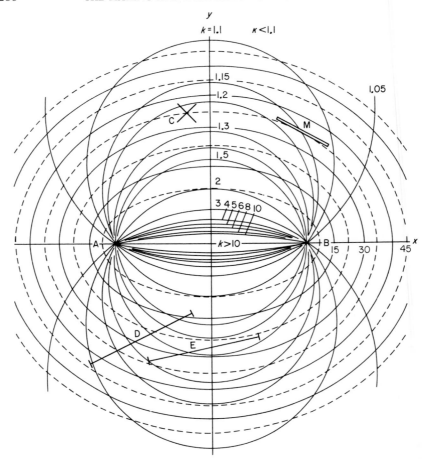

Fig. 7.10. The holo-diagram used to utilize the coherence length. If the object (C) is positioned on the same ellipse as the reference mirror (M) almost no coherence length is needed. If the coherence length of the light is 30 cm, the object at D could not be recorded because the path length difference is more than 60 cm. A slight tilt of the object to the position decreases the difference to less than 30 cm so that a holographic recording can easily be made.

hundred times longer than the coherence length of the beam used. Indeed the train could be as long as the circumference of the ellipse—so long as it followed the path of the ellipse exactly.

Figure 7.10 is drawn such that two adjacent ellipses represent a path length difference of 15 cm when the distance AB is 1 m. Usually the coherence length from an ordinary He–Ne laser is around 30 cm. Therefore, if the reference mirror M is placed at one ellipse, objects can be recorded anywhere within two adjacent elliptical areas at each side of the mirror. This means that if the object is large it may cross four but not five ellipses. Thus the long object at D (Fig. 7.10) cannot be recorded in the hologram, but after a slight change of its position to that of E it can be recorded without difficulty.

Had the distance between the spatial filter and the hologram been 2 m instead, the coherence length of 30 cm would have appeared to be relatively shorter in the diagram. In that case the object may intersect two but not three ellipses; in other words, if the reference mirror is placed at a dotted ellipse, everything in the area limited by the two adjacent ellipses will be recorded. Had no reference mirror been used at all but the reference beam consisted of light going directly from A to B, then in that case everything within the ellipse representing a path length difference of 30 cm would be recorded. In that case, if the separation AB is 2 m, that means everything inside the smallest dotted ellipse of Fig. 7.10.

What has just been described refers to the ellipses of the holo-diagram. However, these ellipses only represent intersections of the axis with the rotationally symmetric ellipsoids of the real world. Thus the coherence space of the preceding example is limited by the ellipsoid representing a maximal path length of 30 cm. In prior examples the useful space is limited by one inner and one outer ellipsoidal shell.

Figure 3.13 shows a pressure vessel with interference fringes representing an expansion. The reason that the image disappears at the left along a vertical line is that the object, in the form of a square box, was so deep that it could not be totally placed within the ellipsoids. Thus the light from its left-hand side, which was the furthest away, had a path length that differed more than 30 cm from that via the reference mirror. As a contrast to this limited recording depth, consider Fig. 7.11, in which is seen a steel bar 2 m long that is bent and twisted by a weight placed at its left-hand side and halfway down its length. The reason why the whole bar could be recorded in spite of its length as compared to the coherence length was that it was placed within the first ellipsoidal shell of the holo-diagram of Fig. 7.10. The author and colleagues have managed to record objects many metres long by using the holo-diagram to optimize the utilization of the coherence length of He–Ne lasers.

7.3.2 Practical examples

When the vertical milling machine of Fig. 6.17 was holographed, the set-up had to be rather carefully planned. The laser used was a 60 mW He–Ne laser with a coherence length of only 30 cm while the machine was close to 2 m high and more than 1 m wide and deep. It would have been possible to position the machine so that the side of the machine being studied could have fitted between two flat surfaces separated by 30 cm. However, the ellipsoids are

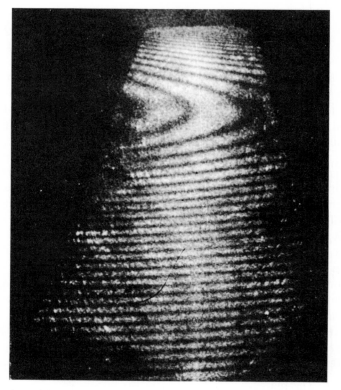

Fig. 7.11. A steel beam that was more than 2 m long has been holographed using a laser with a coherence length of less than 30 cm, using the method of Fig. 7.10. The bar was bent and twisted between the two exposures of the hologram.

curved and it might therefore, at first, appear impossible to make the recording. The solution is to use large distances. The further away one moves from A and B in the diagram the larger is the radius of curvature. If the light source (A) and the hologram plate (B) are close together, the two coherence-limiting ellipsoidal shells can be approximated into spherical shells. If, finally, the distance from the light source to the object is made longer, the radii of the spherical shells will be larger, so that the surfaces of the two limiting shells become more and more flat.

The side of the machine that was to be recorded was almost flat, and at a certain distance it could be positioned within two limiting spheres. When the distance from the light source to the machine was less than about 5 m the reconstructed image of the machine was limited by a dark ring representing the intersection of the machine by one limiting coherence sphere. Thus the distance of 10 m used during the experiments gave a good result in which even cavities and protruding details on the machine could be seen.

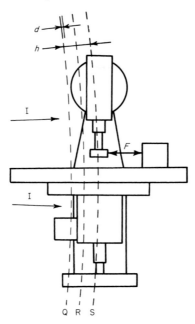

Fig. 7.12. Holographing the almost 2 m high milling machine of Fig. 6.16 using a laser with a coherence length of less than 30 cm. Most of the machine could be positioned between the coherence-limiting ellipsoidal shells Q and S. The displacement needed to produce one fringe is represented by d (which is of course greatly exaggerated). h represents the coherence length. R is the ellipse on to which the reference mirror was placed, while F represents the static load causing the deformation.

The table, however, reached too far out and is therefore not visible in the reconstructed holographic image. Figure 7.12 shows how the machine is positioned in relation to the ellipsoids of limiting coherence length. The divergent laser beam from the spatial filter arrives from the left and, by adjusting the reference mirror, the innermost ellipsoid could be positioned so as just to touch the closest part of the machine. Only the table and an engine casing underneath intersected this ellipsoid and therefore disappeared from the reconstructed image and became black, as seen in Fig. 6.17.

If it had been absolutely necessary to record all the details of the machine the table could have been illuminated separately with light that had travelled a longer way from the laser. It could, for example, have been reflected by one or two mirrors before reaching the table. Another method would be to use two reference beams, each illuminating one half of the hologram plate: if one beam had a shorter path length, it could also be used to record parts of the object that were at shorter distances from the light source.

There exist other methods too, but the one recommended is to use a laser with a longer coherence length, such as an argon laser with an étalon of the type which was used for making the holograms of Figs 6.26–6.33.

7.3.3 The repetitivity of the coherence length

There is one property of the coherence length that has yet to be mentioned. As the path length difference between object beam and reference beam is increased, the image-carrying interference fringes become weaker until they disappear when the difference is equal to the coherence length. If, however, the path length difference is made *still larger*, the fringes will reappear and be just as clear as when this difference was zero. At larger differences again the fringes will again grow weaker and disappear. Thus the coherence phenomenon repeats itself over and over again. The path length difference between maximal coherence will be found to be equal to twice the cavity length $(2L)$ or approximately to twice the length of the laser.

One way to understand this and other coherence phenomena is based on the fact that there exist many different frequencies (axial modes) within the "lasing" spectral line (doppler curve) as shown in Fig. 7.3. When these frequencies are mixed there is a beating effect, the beat frequency being $c/2L$, as verified by the calculations resulting in equation (7.3). Thus the light intensity goes on and off with that frequency, which means that the light waves of maximal intensity travel the distance $2L$ before the next intensity maximum is emitted. When the beam is split into two beams that later recombine at the hologram plate the two components will be of equal intensity at equal time only if the path length difference is either zero or a multiple of twice the cavity length.

The image-carrying interference fringes will, of course, form only if the two beams intersect in time and space at the hologram plate. There will be no fringe-forming process if first the object beam illuminates the plate while the reference beam is absolutely dark and later the reference beam arrives when the object is dark. Therefore the fringes will first be of high quality when the path length difference is zero. As the path length difference is increased, a smaller and smaller portion of the light pulses will illuminate the plate simultaneously until they do not intersect at all. As the path length difference is increased still more, the pulses again begin to arrive at the same time; when the difference is twice the cavity length they will again by synchronized.

This phenomenon can be useful in holography. The holo-diagram can be modified as seen in Fig. 7.13: if the reference mirror is placed at M, any object (C) within the same elliptical area can be recorded. This statement will also be true for objects within every elliptical area that is bright in the figure because they all represent path length differences such that

$$D = n \times 2L \pm l, \tag{7.7}$$

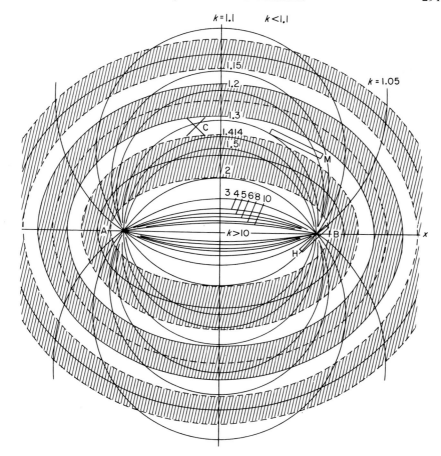

Fig. 7.13. Because of the constant spacing between the different frequencies of the laser light (df of Fig. 7.3), the coherence condition is repeated for every path length difference of double the laser cavity. If therefore a reference mirror is placed in one of the bright elliptical areas, objects in the same or other any other bright area can be holographed.

where D is the path length difference, n is an integer, L is the cavity length (e.g. 60 cm), and l is the coherence length (e.g. 30 cm).

Not only can objects be recorded in all the bright elliptical areas, but also the reference mirror can be moved from one of those areas to another without changing the coherence situation. As the holo-diagram is bright at its centre the holographic set-up would even function in an unchanged manner without any reference mirror at all if the laser light is allowed to pass to the plate directly between A and B.

The object is thought of as illuminated directly by a spatial filter at A. It could, however, just as well be illuminated via mirrors placed in any position such that the path length of the illumination beam via the mirrors is lengthened by twice the cavity length or some multiple thereof. This method can be of value if the coherence length is short and an object needs to be illuminated from more than one direction.

One minor disadvantage with all the methods of utilizing the repetitivity of the coherence length which have been described above is that when the laser is changing its length, e.g. during the warming up process, there will be a varying phase shift between the beams that have travelled different path lengths. Therefore the methods described can still be more problematical than when object and reference beam work within the same coherence area. If a large object is to be holographed, therefore, the author again recommends, if possible, the use of a laser with a long coherence length.

A practical method for the measurement of coherence length is to make a holographic recording of a long object that is so positioned that it intersects as many as possible of the coherence ellipses. The experimental set-up of Fig. 7.14 has been found to be most useful. A 2 m long bar is painted with a retroreflective paint (e.g. 3 M Codit white reflective liquid). The bar is positioned just under the x-axis and to the right of B in Fig. 7.10, where the ellipses are most closely spaced (the k-value of the holo-diagram is close to unity). The spatial filter and the hologram plate are close together at one end of the bar while the reference mirror is fixed to the other, as seen in Fig. 7.14.

When the coherence length is very short only a thin bright line will be seen on the reconstructed image of the bar, close to the reference mirror R. On the

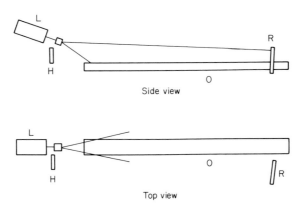

Fig. 7.14. Holographic measurement of the coherence length. The laser (L) illuminates a steel bar (O) that is coated with a retroreflective paint at an oblique angle. The hologram plate (H) is placed close to the laser, while the reference mirror (R) is placed at the other end of the bar. Thus the path length difference between reference beam and object beam will be zero at the far end of the bar and increase closer to the hologram plate.

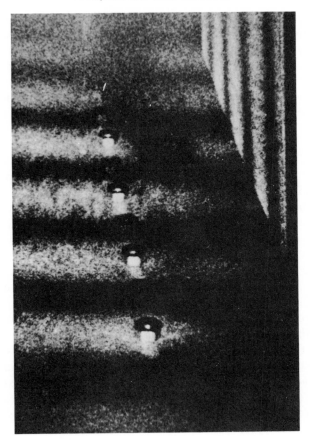

Fig. 7.15. The dark bands in this reconstruction of the bar of Fig. 7.14 represent lack of coherence while the bright bands represent the repetitive coherence function of the light from a singly pulsed ruby laser of low quality.

other hand, if the coherence length is more than 4 m, the whole bar will be reconstructed within a bright area. If there is a repeated coherence length as described in Fig. 7.13, alternating bright and dark areas will be seen on the bar. As an example of this test method Fig. 7.15 represents the result when this procedure is carried out using a ruby laser with a repetitive coherence length of some 10 cm.

7.4 Contouring

Looking at Fig. 7.13 and considering the underlying principles it can be understood that the dark and bright ellipsoids that intersect the reconstructed object can be used to produce depth contours. If the intersecting surfaces had been flat, parallel and equidistant, the dark fringes could be used to reveal the three-dimensional shape of an object: to get a useful result the spacing between the ellipsoids would also have to be much smaller, e.g. a fraction of a millimetre. When all these desired conditions are fulfilled the method works quite well and is referred to in the literature as a contouring method because it produces contours of constant depth.

7.4.1 Two frequencies

Usually the contouring fringes are required to represent depth separations of only a fraction of a millimetre. Therefore the coherence-limiting frequencies have to be further apart than allowed by the doppler curve of the "lasing" spectral line. The easiest way to attain this result is either to expose a hologram using a laser that simultaneously emits light in two different spectral lines or to make a double exposure using a laser the wavelength of which can be changed in between the two exposures.[2] The argon laser and its function has already been described in Figs 7.4 and 7.5; a dye laser is shown in Fig. 7.16.

One way to describe the contouring fringes caused by two frequencies is as a result of the beating effect already mentioned. If one wavelength is 1% longer than the other, it will take 100 wavelengths for the two waves to go from "in phase" to "out of phase" and back to "in phase" again. Thus for every hundred waves there will be constructive interference while in between there will be destructive interference, so that the separation between the dark fringes will represent 100 wavelengths. This relation could be expressed by the following equations:

$$n\lambda_1 = (n+1)\lambda_2 ; \qquad\qquad (7.8)$$

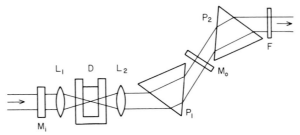

Fig. 7.16. A dye laser, pumped from the left by an argon laser beam which passes through a cavity mirror (M_i), is focused by a lens (L_1) into a dye cell (D) where fluorescent light is emitted. The lens L_2 collimates this light, the prism P_1 is used to select a single colour while the mirror M_o completes the laser cavity. The prism P_2 compensates for the deflection by P_1 and finally an interference filter (F) stops the direct light from the argon laser.

thus

$$n = \lambda_2/(\lambda_1 - \lambda_2) \tag{7.9}$$

and so

$$d = n\lambda_1 = \lambda_1\lambda_2/(\lambda_1 - \lambda_2) = \lambda_1\lambda_2/\Delta\lambda, \tag{7.10}$$

where d is the separation of the depth contours expressed in path lengths, λ_1 and λ_2 are two different wavelengths, $\Delta\lambda = \lambda_1 - \lambda_2$, and n is the number of waves between fringes.

If $\Delta\lambda/\lambda_1$ and $\Delta\lambda/\lambda_2$ are of the order of 10^{-2}, then one contouring fringe will be produced where the object is intersected by every hundred ellipsoids of the holo-diagram. In order to keep the intersecting surfaces (the ellipsoids) flat and equidistant (i.e. of constant k-value) it is essential to work at large distances from A and B or to collimate the illumination and the observation beams.

Thus it has been shown how a lack of coherence length, which in ordinary holography is a disadvantage and a nuisance that has to be overcome, has been turned into a useful method for depth measurement purposes. In the following paragraphs other contouring methods will be described that also utilize the holograms' otherwise unwanted sensitivities to disturbances.

The change in wavelength ($\Delta\lambda$) need not necessarily be caused by a change in the frequency of the laser light, it can just as well be caused by a change in the speed of light (see equation (7.11)). One way to get such a change is by placing the object in a glass cuvette filled with a gas of high refractive index (e.g. acetylene) during the first exposure and another gas with a lower refractive index (e.g. hydrogen) during the second. Now

$$\lambda = nc/v. \tag{7.11}$$

Thus, by reasoning similar to that resulting in equation (7.10), the following result is obtained:

$$d = \lambda n_1 n_2/(n_1 - n_2), \tag{7.12}$$

where d is the separation of depth contours expressed in path lengths, λ is the wavelength of the light, c is the speed of light in a vacuum, v is the frequency of laser light, and n, n_1 and n_2 are the refractive indices in the general case and for the two different gases respectively.

Instead of using two gases, two liquids can also be used, e.g. the "grogg method", in which the object is immersed in a mixture of alcohol and water the concentration of which is changed between the two exposures.[3]

7.4.2 Object translation

When ordinary double-exposure hologram interferometry is used, the light travels from the laser to the hologram plate via the object. If this path length is changed by one wavelength between the two exposures, the interference order at the point studied changes by unity. Thus a particular number of fringes on the reconstructed object represents a particular difference in path length over the object area. The reason for this difference in path length may

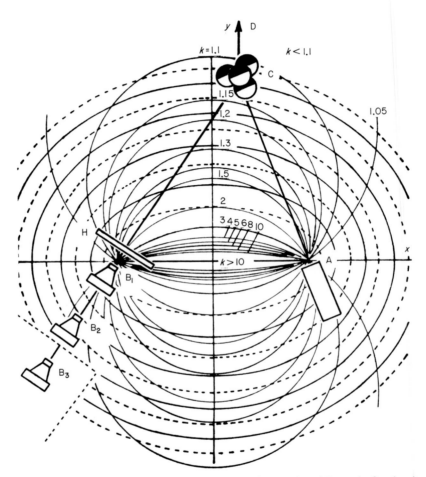

Fig. 7.17. The holographic set-up used in the experiment. A and B are the focal points of the ellipses of the holo-diagram. The object (C) consists of four white-painted table-tennis balls placed in the form of a pyramid, which is illuminated by divergent light from a laser at A. Dark parts of the balls represent shadows. The holographic plate (H) is placed close to B_1 during exposure. Between the two exposures of the hologram C was displaced in the direction of the arrow D. During reconstruction C was photographed from B_1, B_2 and B_3, all of which lie along the same line of sight towards C. The reference beam and the identical reconstruction beam are excluded from the figure. Distances: $AB_1 = 720$ mm; $BC_1 = 1125$ mm; $CA = 1014$ mm; $B_1B_2 = 250$ mm; $B_2B_3 = 2000$ mm.

be either that different parts of the object have moved different distances (rotation or deformation) or that a rigid body translation influences the path length differently at different parts of the object, or a combination of these two effects. If a large displacement is needed to cause a certain path length difference, the desensitivity factor (k) of the holo-diagram (Fig. 7.17) is high. Rigid body translations of the object will cause fringes if the k-value varies over the surface of the object. These fringes represent the intersection of the object by a set of interference surfaces in space.

The three-dimensional shape of these interference surfaces can be visualized by using the moiré pattern analogy to hologram interferometry that was introduced in Section 5.5. A holo-diagram similar to that of Fig. 1.24(a) was drawn but with more closely spaced ellipses. Every second spacing between the ellipses was painted black. Thus the thickness of the black and the white stripes is proportional to the k-value. A is the spatial filter, B is an observation point just behind the centre of the holographic plate (H), C is the object which, between the exposures, was moved in the direction of the arrow D. The resulting interference fringes are found by making a transparent copy of the set of ellipses, placing it exactly on top of the original ellipses and finally

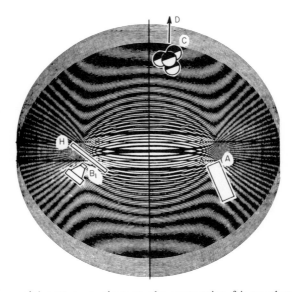

Fig. 7.18. The moiré pattern analogy to the contouring fringes that are recorded when the camera B_1 photographed the reconstructed object. The intersection of the object (C) by the moiré fringes caused by the displaced elliptic pattern represents the intersection of the real object by imaginary interference surfaces in space, caused by holographic double exposure. The directions, shapes, and relative spacings of the moiré and interference fringes, respectively, are analogous, but the absolute value of the spacing has to be calculated.

displacing the transparency in the direction of the arrow D. The result is found in Fig. 7.18, where a set of ellipses was used that was still more closely spaced than those of Fig. 1.24(a). The moiré fringes of Fig. 7.18 are analogous to the interference surfaces of the holographic experiment. However, they represent just one two-dimensional cross-section (but a very important one) of the three-dimensional interference surfaces. To produce the interference fringe surfaces in space *in toto*, imagine the three-dimensional moiré pattern of two sets of rotationally symmetric ellipsoids by studying its most characteristic cross-section. From Fig. 7.18 it is found that the interference surfaces in the vicinity of C are surprisingly flat and that they intersect the line of sight (radius from B) at an angle that differs from zero.

This suggests that contouring fringes can be produced by a simple translation.[4] A set-up similar to that illustrated in Fig. 7.17 ought to be able to produce contouring fringes that are perpendicular to the line of sight and that also represent intersecting surfaces that are more flat than would be expected with regard to the curvature of the ellipsoids or the wavefronts.

One further advantage would be that a motion of the point of observation backward and away from the plate during reconstruction will influence the moiré pattern of Fig. 7.18 so that the interference surfaces rotate and, at a

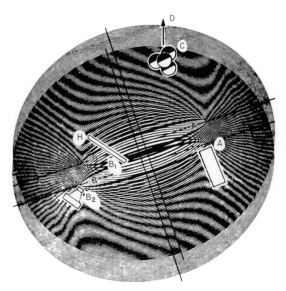

Fig. 7.19. When the camera is moved backwards from B_1 to B_2 during reconstruction the focal point B of the elliptical pattern should follow while the other focal point A remains fixed to the laser. As before one pattern is thereafter displaced in the D direction, and the moiré fringes that form represent the holographic contouring fringes as photographed from B_2. These fringes are now perpendicular to the line of sight (B_2C).

Fig. 7.20. Photograph taken with the camera at B_1 of Fig. 7.17. Observe the similarity to the pattern one would see from B_1 of Fig. 7.18 if the object were intersected by the moiré fringes. To reduce speckles this photograph was made using the comb method.

certain distance, become perpendicular to the line of sight. This is demonstrated in Fig. 7.19, which was made in the following way. The first set of ellipses was placed with one focus (A) at the spatial filter, as when Fig. 7.18 was made. The other focal point, however, was placed at B_2 instead of B_1 (see Figs 7.17 and 7.19). The moiré pattern was once more produced by displacing a transparent copy of the ellipses in the direction of the arrow D.

Some practical holographic results are demonstrated by Figs 7.20, 7.21 and 7.22. It was found that the practical results were in qualitative agreement with the visual experiments using the moiré pattern analogy.

Hologram interferometry is often used for non-destructive testing. For that purpose it is important to know that a complicated fringe pattern can be caused not only by object deformation but also by contouring of an undeformed, translated object having a complicated shape. By studying the fringe patterns on table-tennis balls that are glued to the object, it is possible to determine whether contouring fringes have formed and what type of motion caused them.

It is still too early to say whether the described contouring method can be recommended for practical measurement purposes. One important drawback is that the object has to be physically displaced by a relatively large amount

Fig. 7.21. The same reconstruction as that of Fig. 7.20, but the camera was placed at B_1 of Fig. 7.17. The intersecting interference fringes are perpendicular to the line of sight (B_2C), as are the moiré fringes of the analogous experiment of Fig. 7.19. To get sharp fringes this photograph was recorded with a vertical slit in front of the camera lens.

while its sensitivity to tilt and deformation is some hundred times higher than the sensitivity to contouring. There are many reasons to believe that the method using sandwich holography is superior because it makes possible independent displacements of A and B without displacement of the object.

7.4.3 Combing the speckles out of the fringes

If the translation of the object is large the fringes will move very rapidly when the observer moves his head along the hologram plate during reconstruction. If the observer wants to photograph the fringes he has to use such a small aperture that the entire utilized area of the camera lens "sees" the fringes at the same position. Otherwise the fringes will be blurred out in the photograph. Sometimes, but not always, the situation can be improved by focusing the fringes and defocusing the object (see Section 5.3).

As a result the following simple approach was developed[5] and the trouble was literally combed out! An ordinary comb is placed in the reconstruction beam so that it throws a shadow on the holographic plate. By tilting the comb and by moving it along the divergent beam the direction and the

Fig. 7.22. The same reconstruction as that of Fig. 7.19, but moving the camera further away to B_3 of Fig. 7.17 changes the direction of the intersecting interference surfaces still more. This photograph was recorded using a rather small aperture.

separation of the shadows of the teeth can be changed until it is found that, whatever illuminated area of the plate the observer looks through, the fringe pattern of the reconstructed object is the same. (The separation of the shadows of the teeth is now the same as would be the separation of Young's fringes if one point on the object could simultaneously illuminate the plate from its positions before and after displacement.) The comb is fixed in this position, and the camera will produce a clear photograph of the reconstructed object with its fringes (Fig. 7.20).

However, when the author tried to use this new comb method while recording the different contouring patterns of Figs 7.20, 7.21 and 7.22 it was found, much to the surprise of the author and his colleagues, that all the photographs became identical to Fig. 7.20, independent of the distance between the plate and the camera. Thus the rotation of the contouring intersecting interference fringe surfaces appears to be caused by the same parallax effect as is ordinary fringe motion.

7.4.4 Illumination and observation shearing

There are many methods for the production of interference fringes that will represent the intersections of a three-dimensional object by a set of equally spaced, virtually plane surfaces. These fringes, contouring fringes, reveal the topography of the object as do the contour lines of a map. In Chapter 1, it was shown that there are three methods of producing interferometric information: (a) by using two coherent points of illumination; (b) by using one point of illumination and one coherent point of observation; and (c) by using two coherent points of observation.

The fringes formed can all be represented by the moiré pattern in space of two sets of spherical wavefronts, their centres being the two coherent points. If the two sets of spherical wavefronts both move outward (or both move inward), a set of hyperboloids is formed (their foci being the centres of the spherical wavefronts). This is the case with methods (a) and (c). If one set of spherical wavefronts moves outward while the other set moves inward, the interference fringes form a set of rotational ellipsoids (their foci being the centres of spherical wavefronts). This is the case with method (b) (cf. Section 1.4).

Let us consider how these three methods can be used for holographic contouring.

(a) Two coherent points of illumination

The hologram is doubly exposed, and the point source illuminating the object is moved slightly sideways between the two exposures. The reconstructed holographic image reveals the object intersected by surfaces that consist of a set of rotationally symmetric hyperboloids (their common foci being the two locations of the point of illumination). Thus the fringe pattern is independent of the point of observation. (In accordance with Gabor's "subjective and objective speckles" the author would like to introduce the term "objective fringes" for this type of fringe. Fringes that depend on the point of observation would consequently be named "subjective fringes".) Exactly the same result would be produced if holography was not used at all, but rather the object was examined directly when it is illuminated by the two points simultaneously. (This type of fringe is often referred to in the literature as a "projected interference fringe".) The longer the distance between object and source of illumination the more flat are the intersecting surfaces that are approximately parallel to the illuminating beams. Collimated beams produce equidistant, parallel, flat surfaces.

The main disadvantage of using this method for contouring purposes is that its sensitivity is zero if the direction of illumination and the direction of observation coincide. Usually the intersecting surfaces are deliberately arranged to be normal to the line of sight (as is the case with the contour lines of maps). To produce this result the illuminating beam also has to be normal to the line of sight, and in that case much of the object surface would be hidden by shadows. The main advantage of the method is its simplicity.

(*b*) *One point of illumination combined with one coherent point of observation*
In contrast to the first method this is a true holographic method in which the hologram is doubly exposed. Between the two exposures the wavelength is changed slightly by altering either its frequency or its velocity as described in Section 7.4.1. If a double-frequency laser is used, the two exposures can be made simultaneously. The holographic image reveals the object intersected by surfaces that consist of a set of rotationally symmetric ellipsoids (their common foci being the point of illumination and the point of observation, respectively). Thus, the fringe pattern is dependent on the point of observation (subjective fringes). Exactly the same result would be produced if holography was not used at all, but rather the object was examined using an ordinary Michelson interferometer illuminated by a single wavelength (λ_3) such that

$$\lambda_3 = \lambda_1 \lambda_2 / (\lambda_1 - \lambda_2), \tag{7.13}$$

where λ_1 and λ_2 are the two wavelengths used in the contouring methods described earlier.

The intersecting surfaces are normal to the bisector of the angle separating the beams of illumination and observation. If the distance between the object and the nearest focal point is large, the intersecting surfaces are approximately flat. Collimated illumination and observation beams produce equidistant, parallel, flat surfaces. The main advantage of this method is that the intersection planes are normal to the line of sight if the direction of illumination and the direction of observation coincide. The main disadvantage of the method is its complexity because either a two-frequency laser is needed or the refractive index of the medium (gas or liquid) surrounding the object has to be changed between the two exposures (see Section 7.4.1).

(*c*) *Two coherent points of observation*
A hologram is doubly exposed. The hologram plate is moved slightly between the two exposures. The reconstructed holographic image reveals the object intersected by surfaces that consist of a set of rotationally symmetric hyperboloids (their common foci being the two positions of the point of the hologram plate that is used during observation). Thus the fringes depend on the points of observation (subjective fringes). Exactly the same result would be produced if holography was not used at all, but rather one object was examined directly, straight through a conventional interferometer (see Fig. 1.18). The object that has to be illuminated by a coherent light source is thus simultaneously studied from two observation points that are slightly displaced. Similar results can even be seen through a telescope, the objective lens of which is blocked but for two small holes diametrically opposed to each other.

The larger the distance between object and observation, the more flat are the intersecting surfaces that are (approximately) parallel to the line of sight. Collimated observations produce equidistant, parallel, flat surfaces.

The described method is, however, useless for contouring purposes because the intersecting surfaces are parallel to the line of sight. The fringes thus appear as straight lines that are totally independent of the topography of the object.

7.4.5 Sandwich hologram

In the previous chapter (Chapter 6), it was described how sandwich hologram interferometry makes possible a manipulation of the fringes during reconstruction. When sandwich holography is applied to contouring it thus produces similar advantages in this field.[6] Sandwich holography can replace double-exposure holography whatever contouring method is used. Most is gained, however, when applied to method (a) (a minute displacement of the point of illumination between the two exposures). Not only does such a procedure succeed in combining the best feature of that method (its simplicity) with the advantage of the two-wavelength method (intersecting planes that are normal to the line of sight), but also the possibility of changing the angle of the intersecting planes by more than 90° in any direction of the three-dimensional space during the reconstruction is realized. The experiment was carried out by the author in the following way (Fig. 7.23).

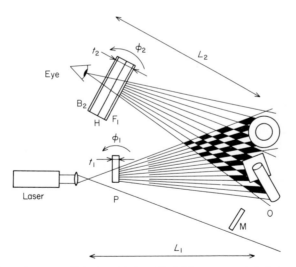

Fig. 7.23. P is a plane parallel glass plate with thickness t_1 rotated through an angle ϕ_1 between the two exposures. H is a sandwich hologram, its two plates B_2 and F_1 representing second and first exposures, respectively. The two plates are bonded with their emulsions outward. The sandwich with thickness t_2 is rotated through an angle ϕ_2 during reconstruction. L_1 and L_2 are distances from objects O to the points of illumination and observation, respectively. The chains of rhombs represent the contouring fringes.

A holographic plate with its emulsion stripped away was placed in a plate-holder behind an ordinary holographic plate (F_1) that had its emulsion directed toward the objects (O). The distance (L_2) from the plate-holder to the objects was some 55 cm. Close to the three objects (O) was placed a mirror (M) that reflected the reference beam towards the hologram. The object and the mirror were illuminated by divergent light from a spatial filter 100 cm away. The angle between the directions of illumination and observation was some 30°. Close in front of the spatial filter was placed a plane parallel glass plate (P) which was 11 mm thick. A rotation of this glass plate through the angle ϕ_1 produced a lateral displacement of the beam illuminating the object.

After the first exposure the glass was rotated through some 2°. The plate-holder was emptied, and a new holographic plate (B_2) was placed in the plate-holder having its emulsion directed away from the object. The holographic plate with its emulsion stripped away was placed in front, and a second exposure was made.

The holographic plates from the two exposures were processed and repositioned in the plate-holder. The first exposed plate was placed in

Fig. 7.24. The reconstruction beam illuminates the sandwich hologram from approximately the same direction as did the reference beam during exposure. The contouring fringes are identical to the projected fringes of Fig. 7.23 that radiate from the focus of the lens in front of the laser. The intersecting surfaces consist of a set of hyperboloids (their common foci being the apparent points of illumination during the first and the second exposures).

Fig. 7.25. The same sandwich hologram as that in Fig. 7.24 but during reconstruction the hologram was tilted with respect to the reconstruction beam. This way the intersecting interference surfaces could be rotated until they become perpendicular to the line of sight (see the moiré pattern in Fig. 7.23).

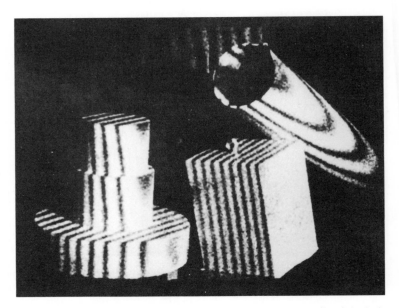

Fig. 7.26. The same sandwich hologram as that in Fig. 7.24 but a tilt of the hologram during reconstruction has rotated the intersecting surfaces so that they become parallel to one side of the cube.

Fig. 7.27. The same sandwich as that in Fig. 7.26 but the intersecting surfaces have been rotated through 90° during reconstruction so that they become parallel to the other side of the cube.

Fig. 7.28. The same sandwich hologram as that in Fig. 7.27. The hologram was tilted around an almost horizontal axis during reconstruction. That way the intersecting surfaces were rotated through 90° so that they become parallel to the top side of the cube.

front of the one exposed second so that both plates of the sandwich hologram had their emulsions outward (Fig. 7.23). When illuminated by the beam from the reference mirror, the reconstructed objects appeared covered by fringes that behaved just as any contouring fringes produced by two-point illumination. Thus the fringes appeared fixed to the object surface and directed toward the illumination source (parallel to the shadows). If, however, the two plates were sheared in relation to each other, the direction of the fringes changed.

The two plates were bonded together while their positions were fixed by the plate-holder. The cement used (which had the trade name Cyanolite) hardened in less than 10 s. The sandwich hologram thus formed could be reconstructed while being hand-held, and when it was tilted the fringe pattern changed. In that way it was possible to tilt the intersecting interference surfaces through more than 90° in any direction of the three-dimensional reconstructed object space. To demonstrate this possibility Figs 7.24–7.28 are all photographs of reconstructions of a single sandwich hologram. Figure 7.24 shows the original fringe system (no sandwich tilt). In Fig. 7.25 the intersecting planes are perpendicular to the line of sight. In Figs 7.26–7.28 the sandwich hologram has been tilted so that the intersecting planes are parallel to each of the three sides of the cube.

The two exposures represent two coherent and simultaneous points of illumination that produce a fringe system (Young's fringes) in space in the form of a set of rotational hyperboloids, their common foci being the two points of illumination separated by the distance h_1. The distance d between two adjacent hyperboloid shells is

$$d = \lambda/2 \sin \alpha, \qquad (7.14)$$

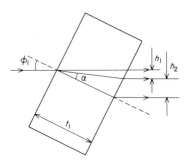

Fig. 7.29. This figure is used to demonstrate two different situations: (a) illumination shear (h_1) caused by a rotation (ϕ_1) of a plane parallel glass plate (of thickness t_1 and refractive index n_1); (b) observation shear (h_2) caused by a rotation (ϕ_2) of a sandwich hologram (in which the thickness of glass separating the two emulsions is t_2 and refractive index is n_2).

where λ is the wavelength of light and 2α is the angle between the directions of the two points of illumination as seen from the object.

Referring to Fig. 7.29, the shear caused by the rotation of the glass plate is given by

$$h_1 = t_1 \left[\sin \phi_1 - \frac{\sin 2\phi_1}{2(n^2 - \sin^2 \phi_1)^{1/2}} \right], \qquad (7.15)$$

where h_1 is the distance through which the illumination beam has been sheared, t_1 is the thickness of the glass plate P, ϕ_1 is the angle of rotation of P, and n_1 is the refractive index of P. Approximating to small ϕ_1,

$$h_1 = (t_1 \phi_1 / n_1)(n_1 - 1). \qquad (7.16)$$

Thus the separation (d) of the illumination fringes is found by combining equations (7.14) and (7.16) and by substituting $h/2L$ for α:

$$d = n_1 \lambda L_1 / \phi_1 t_1 (n_1 - 1), \qquad (7.17)$$

where L_1 is the distance from the illumination point to the object (Fig. 7.23).

A tilt of the sandwich hologram during reconstruction is analogous to a shearing of the observation system. Consequently, it can be represented by two coherent simultaneous points of observation that produce an imaginary fringe system in the form of a set of rotationally symmetrical hyperboloids, their foci being the two points of observation. The distance (x) between two adjacent hyperboloid shells is

$$x = \lambda / (2 \sin \gamma), \qquad (7.18)$$

where 2γ is the angle between two points of observation as seen from the object. For the shearing effect of sandwich hologram tilt refer to equation (3.9).

Referring to Figs 7.23 and 7.29, the shear of the observation beam is given by

$$h_2 = \frac{t_2 \sin 2\phi_2}{2(n_2^2 - \sin^2 \phi_2)^{1/2}} \qquad (7.19)$$

where h_2 is the distance through which the observation beam has been sheared by rotation of the sandwich hologram (H of Fig. 7.23), t_2 is the total thickness of the glass plates between the two emulsions, ϕ_2 is the angle of rotation of H, and n_2 is the refractive index of the glass plates of H. Approximating to small ϕ_2,

$$h_2 = t_2 \phi_2 / n_2. \qquad (7.20)$$

Thus the separation (x) of the observation fringes is found by combining equations (7.18) and (7.20) and by substituting $h_2 / 2L_2$ for γ:

$$x = n_2 \lambda L_2 / \phi_2 t_2, \qquad (7.21)$$

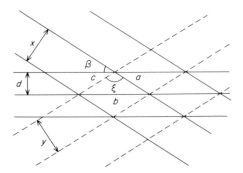

Fig. 7.30. Study of the contouring fringes caused by illumination and observation fringes (indicated by rows of rhombs in Fig. 7.23). Illumination and observation are made from the left of the figure.

Table 7.1   Relationships

y	ξ	x
0	0 or π	± 0
d	β or $\pi - \beta$	$d/(2\cos\beta)$ or ∞
$d/(\sin\beta)$	$\pi/2$	$d/(\cos\beta)$

where L_2 is the distance from the point of observation to the reconstructed object (Fig. 7.23).

The three-dimensional coherent moiré pattern of these two sets of hyperboloid interference fringe surfaces (caused by two points of illumination and two points of observation, respectively, as described in equations (7.17) and (7.21)) produce a third three-dimensional interference fringe system (Fig. 7.23). The intersections of the object by these resulting fringes produce the contouring fringes seen in the reconstruction.

In the following paragraphs mutual interference of the illumination fringes and the observation fringes to produce contouring fringes will be studied mathematically. Assume that the distances (L_1 and L_2 of Fig. 7.23) are so large that the fringe surfaces within the volume under scrutiny can be approximated by flat, equally spaced, parallel planes.

The following equations refer to Fig. 7.30:

$$d = a\sin\beta, \tag{7.22}$$

$$x = b\sin\beta, \tag{7.23}$$

$$c^2 = a^2 + b^2 - 2ab\cos\beta, \tag{7.24}$$

$$cy = bd_1, \tag{7.25}$$

$$\sin\xi = (y/d_1)\sin\beta. \tag{7.26}$$

The combination of equations (7.22)–(7.25) results in

$$y = \frac{xd_1}{(d_1^2 + x^2 - 2d_1 x \cos \beta)^{1/2}}. \qquad (7.27)$$

Some relations between x, y and ξ based on equations (7.23) and (7.25) are given in Table 7.1.

The fringe distance (y) is at a maximum when the fringes are perpendicular to the line of sight. Theoretically the fringes can be rotated through up to 180°. However, low fringe contrast caused by decorrelation sometimes limits the useful value of fringe rotation. Repeated experiments have shown that the fringe contrast was highest if the plane parallel glass plate (P) was rotated in the direction of the arrow as shown in Fig. 7.23.

For measuring and demonstration purposes sandwich contouring holograms present many unique possibilities, one of which being that the direction of slant on the object (uphill or downhill) is easily found. It can even be arranged that the angle of tilt of the sandwich hologram corresponding to fringe-free reconstruction has the same direction as the angle of slant of the object. The relationship between the values of the two angles is found from equations (7.21), (7.26) and (7.27).

The rear plate of a sandwich hologram is both exposed and reconstructed through a front plate. If all plates were identical there would be no difference in optical path length when the front plate is exchanged. The effect of differences in thickness and alignment of the glass plates is to a large degree compensated for by the fact that the virtual image from the front plate is seen through approximately the same area of the plate, as the reconstruction beam has to pass through to illuminate the rear plate. Thus only local variations in thickness have any appreciable aberration effects that can cause false fringes. When the sandwich hologram is studied from a distance of only a few centimetres, the error is usually less than a quarter of a fringe.

If higher accuracy is needed, special optically flat plates can be substituted for the ordinary glass base of the hologram plates. It is also recommended that the equations presented in this chapter be used only for guidance. Calibration techniques should be employed whenever possible. Reference objects with known dimensions and angles ought to be placed near the object.

From a theoretical point of view this new contouring method is very interesting because it appears to be the first example in the literature of a combination of illumination and observation shearing. The following methods (introduced in Chapter 1) assist in a very practical way in the understanding of this type of phenomenon.

Two coherent points of illumination produce two sets of spherical wavefronts. The three-dimensional moiré effect of these intersecting wavefronts is interference surfaces in space in the form of a set of rotationally symmetric hyperboloids. The common foci of these surfaces are the two points of illumination. This kind of fringe is well known as Young's fringes.

Two coherent points of observation produce interference fringe surfaces in space of similar type. The two points of observation represent the common foci of a set of rotationally symmetric hyperboloids. However, these fringes are not real; they exist only when defined by an observer. Thus they could be called subjective fringes.

The coherent three-dimensional moiré pattern of these two sets of rotationally symmetric hyperboloids produces a third set of interference fringe surfaces in space. A coherent moiré pattern is a moiré pattern that produces the same result as interferometry (in which constructive as well as destructive interference can occur). Finally, the intersection of the three-dimensional object by the third set of interference surfaces produces the contouring fringe seen in the reconstructed object space. These fringes represent the difference between object shape and the shape of the fringe surfaces.

7.4.6 Practical considerations

All the methods of producing contouring fringes described above can easily occur accidentally during an experiment: the laser frequency might change (with the exception of the frequency of ordinary He–Ne lasers); the illumination source, the hologram plate or the object itself could change position; the temperature of the object or the temperature of the atmosphere might be altered.

Thus, in the author's opinion, the contouring principles are most important as possible causes of error when hologram interferometry is used for measurement of deformation.

One way to check whether unwanted contouring has occurred is, of course, to see whether the fringes under examination appear to be influenced by the shape of the object. It is, however, sometimes very difficult to get a definite answer in this way as the deformation itself usually depends on the object shape as well. Therefore it is recommended that spheres or cones be bonded to the object, the cones with their points towards the hologram plate. If there are only straight fringes across these test bodies no contouring has occurred, but if there are curved or circular fringes these indicate that there has been a contouring process that most probably has influenced the whole object.

There are two reasons why holographic contouring methods are not recommended for measurements. One is that in all the described techniques the depth resolution is lowered without lowering the sensitivity to errors. If, for example, every contouring fringe represents a depth of 100λ, the sensitivity to changes in depth between the two exposures might still be 0.5λ. Thus a deformation of, say, $0.3\,\mu m$ would result in an error in measured depth of no less than $30\,\mu m$.

The second reason is that there exist some very good non-holographic contouring methods, e.g. the shadow moiré technique. In this technique a grid is placed close to the object so that both illumination and observation are made through the grid. When illuminated by a point source the grid throws shadows on to the object just as the Young's fringes are projected in Fig. 7.23. Observation through the grid produces the same result as the Young's fringes of observation and thus the object will be seen intersected by depth surfaces. In this case the fringes of blindness (Section 5.5.7) are simply caused by the grid lines obscuring the field of view. Thus, in Fig. 7.23, the plane parallel glass plate (P) and also the whole sandwich hologram (H) can be removed. Instead a grid is inserted between the letters L_1 and L_2. If the light source and the point of observation are at the same distance from the grid, the intersecting surfaces will be flat and parallel to the grid. They would also be equidistant if the illumination and observation beams were collimated or were of infinite length.

7.5 The analogy between short coherence length and short pulse length

In Section 7.1.2 coherence phenomena were explained as resulting from the beating effect of different light frequencies. This analogy between a short coherence length and a short light pulse is very useful. It represents more than a mere analogy as, whatever the light source, it can never emit a pulse of light with a coherence length longer than the pulse length. Thus, replacing the coherence length by a pulse of light with the same length will simplify the visualization of coherence effects. Previously in this text this method has been used in cases where there has been no doubt that the coherence phenomena were caused by light pulses. Now this analogy will be stretched a little further.

From an interferometric point of view a short pulse length represents the ultimate example of a short coherence length as there can never be any doubt that no coherence exists outside the pulse. Let us see if this analogy can assist understanding of how ordinary optical phenomena depend on the coherence effect.

7.5.1 Coherence requirements of optical components

A single short light pulse will not cause interference fringes in an interferometer where the two arms have such a path length difference that the two pulses do not intersect in space and time. Thus an interferometer depends on the coherence of the light, which has already been dealt with in Chapter 1.

Let the described interferometer be so adjusted that for continuous light of high coherence the output would be zero because of destructive interference. The difference in arm length, however, results in the fact that only a very short pulse or a train of short pulses could pass through that interferometer if they do not intersect in space and time. When a longer pulse is sent through the interferometer its beginning and its end would pass through while the middle part would be cut out by destructive interference. Thus the interferometer would work like a differential filter which only lets through the variations in intensity.

Is there any need for temporal coherence when an elliptical mirror is used for focusing? If a short light pulse starts at one focal point A, it will be reflected by the ellipsoid and reach the other focal point B (see the holo-diagram of Fig. 1.24(a)). As the distance from A to B via the ellipsoid is constant by definition, all the rays will arrive in phase and thus produce the diffraction-limited point at B by interference. No coherence is needed as there is no time delay between different rays. The same reasoning applies to parabolic, hyperbolic and plane mirrors.

Is there any need for a coherence when a Fresnel zone lens is used to focus light? The lens can be visualized as a cross-section of the ellipsoids of the holo-diagram (Fig. 1.24(a)). Different rays arrive from A to B via different ellipsoids and thus the path lengths differ. The light pulse will therefore be split into many pulses arriving in B at different time intervals. They therefore cannot interfere destructively or constructively at B to build up a point of light if the coherence length is short. Thus a Fresnel zone lens depends on coherence.

Does an ordinary glass lens depend on the coherence of light? The function of an ordinary lens is such that all the rays from a point source are delayed by the same amount by the refraction of the glass. Thus a diffraction-limited point image is formed without the need for a high temporal coherence.

Does deflection by a prism depend on temporal coherence? The answer is no. However short a light pulse is, its shape will not be changed by the prism.

Does the effect of a diffraction grating (e.g. as used in a spectroscope) depend on the temporal coherence? The answer must be in the affirmative. If the pulse was shorter than the spacing of the grating, the light would not be diffracted in any preferred direction at all. If the pulse is much longer than the line spacing but much shorter than the total area of the diffraction grating, then the optical set-up should be so arranged that all diffracted parts of the pulse arrive simultaneously at the detector. As a diffraction grating represents a cross-section of the ellipsoids, it is impossible to produce a grating for which all the rays from A to B have equal path length.

The closest that has been reached to this goal is by the use of a concave grating in a configuration known as a Rowland circle. It can be proved that this circle is identical to a k-circle of the holo-diagram. Let us therefore re-examine Fig. 4.19, where A is the light source, e.g. the slit of the spectrometer. The grating is ruled on the surface of an ellipse close to the y-axis, e.g. where the k-value of 1.15 is printed. The photographic film used for recording is curved so that it coincides with the k-circle with the value $k = 1.15$ and is positioned close to B.

The analogy between coherence length and pulse length has been used above to visualize some coherence functions. In the following section the problem will be turned the other way around and it will be explained how a short coherence length can be used to demonstrate the function of short pulses. This new method of visualizing light in flight can be used to verify the statements given above.

7.5.2 Light-in-flight recording by holography

Light-in-flight recordings using two photon fluorescence[7] or ultrafast Kerr cells driven by laser pulses have been possible for some 10 years.[8] Very fast optical phenomena, such as refractive index changes in laser-produced plasmas, have been recorded by holography using short illumination pulses.[9] However, it appears that the inherent properties of holography have never been used directly to produce a frameless motion picture of ultrafast phenomena, such as light in flight. The possibility of such an experiment was noted in 1972 in a paper describing the properties of the holo-diagram.[10]

In a hologram only those parts of an object will be recorded for which the path length from the laser to the holographic plate via the object does not differ from the path length of the reference beam by more than the temporal

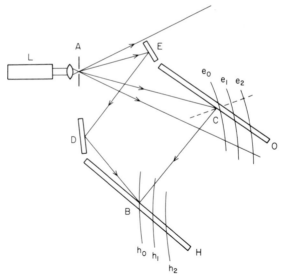

Fig. 7.31. L, laser; A, spatial filter; O, object; C, observed point; H, hologram plate; B, point of observation; and D and E, two mirrors. e_0, e_1 and e_2 are portions of ellipsoids perpendicular to the bisector of the angle ACB, while h_0, h_2 are portions of hyperboloids parallel to the bisector of CBD. From the intersection of H by h_0, only those parts on H that represent its intersection by e_0 are seen because the path length ACB is equal to AEDB.

coherence length of the laser light used for the recording. If the coherence length is short, during reconstruction a large object will be seen intersected by one bright fringe of near-zero path length difference. This fringe represents the object intersected by an imaginary interference surface in the form of an ellipsoid, one of its two foci being the point from which the spherical wavefronts of illumination are emitted (the spatial filter A of Fig. 7.31) and the other focus being the point on the holographic plate used for the observation (B). Thus, for each point on the plate there exists a corresponding ellipsoid in the image space representing zero path length difference between object and reference beams.

When the point of observation is moved around along the holographic plate in a random way, the bright fringe of zero path length will move around on the object. However, the same point on the object will be within the bright fringe all the time if the point of observation (B) is moved along the intersection of the plate and a hyperboloid, its two foci being the studied point at the object (C) and the point source (A) of the reference beam or its mirror image. Thus, for each point on the object there exists a corresponding hyperboloid in the space of the observation representing zero path length difference between reference and object beams.

The motion, parallax, and apparent localization of the bright fringe of zero path length differences differ totally from those of ordinary holographic interference fringes, which are not influenced by the angle or divergence of the reference beam.

By careful planning it is possible to optimize the geometry of the holographic set-up in such a way that the utilized parts of the ellipsoids and the hyperboloids can be approximated to flat surfaces. In that case, the object is seen intersected by one flat imaginary interference surface, the depth position of which can be altered during reconstruction by changing the point of observation at the hologram plate.

Thus, the light of short pulse duration or short coherence length can be used in a new holographic method for contouring. If, however, the three-dimensional shape of the object is known a priori, then this new method could be used instead to study the shape of the imaginary interference surfaces.[11] If the distance between object and hologram plate is long and if the object is a flat surface perpendicular to the line of sight, then the intersections of the ellipsoids can be approximated into intersections of the spherical wavefronts of illumination.

Thus, light of short pulse duration or short coherence length can be used to study the shape of wavefronts by recording a cross-section of the light in flight. By moving the point of observation so that it crosses a number of hyperboloids, it is possible to study the object at different time situations; by moving the point of observation continuously, it is possible to study the light in flight as a continuous high-speed motion picture. This type of holography could be compared to gated viewing, where the gating effect is caused by the fact that only light beams of the same path length will recognize one another and interfere with one another to produce the image-carrying primary fringes on the holographic plate. As the inclined reference beam travels along the plate, this gating effect accompanies it, so that one side will react to "younger" object light than the other.

The relation between space and time is represented in object space by the separation of the ellipsoids (d_{ell}), while the corresponding relation in the observation space is represented by the separation of the hyperboloids (d_{hyp}) in the following way: $d_{ell} = kc/2$ and $d_{hyp} = k^*c/2$, where k and k^* are constants from the holo-diagram ($k = 1/\cos \alpha$ and $k^* = 1/\sin \beta$, where 2α is the angle between the point of illumination and the point of observation as seen from the studied object point, i.e. $2\alpha = $ OCB, and 2β is the angle between object beam and reference beam at the hologram plate, (i.e. $2\beta = $ CBD) and c is the speed of light.

In the present experiment an opaque, white-painted, diffusely flat surface (O) was illuminated from the left at an oblique angle (Fig. 7.31). The path length of the object beam travelling from the spatial filter (A) in front of the laser (L) to the middle (B) of the hologram plate (H) via the middle (C) of the object surface (O) was equal to the path length of the reference beam travelling from the laser to the middle of the hologram plate via two mirrors (D and E). The plate was illuminated from the left at an oblique angle by the reference beam. Thus the left-hand part of the hologram plate was sensitive during recording only to young object light, the age of the light being the time of flight from the laser.

A conventional Spectra-Physics Model 170-03 ion laser was used as light source, the coherence length of which was shortened by taking out the étalon while leaving the prism for wavelength selection in place. All strong wavelengths were tested and gave almost identical results. The object area was about 1 m × 1 m, and the exposure time was 12 s with an output power of 2 W. The recording medium was a conventional Agfa-Gevaert Holotest 8 E 56 plate with an area of 30.3 cm × 25.4 cm. The object beam was slightly stronger than the reference beam.

Light that originates from a laser (without étalon) with a cavity length exceeding 10 cm contains a number of separate bands of frequencies (axial modes). When these different frequencies are mixed, a beating effect results, in that the light becomes intensity modulated. Looking at the coherence phenomenon in this way, it is quite clear that the image-forming fringes on the holographic plate can only form when the object beam and the reference beam illuminate the plate from the same place at the same time. The fringe can, of course, not form if the path length difference is such that one component of the light has maximum intensity while the other is completely dark.

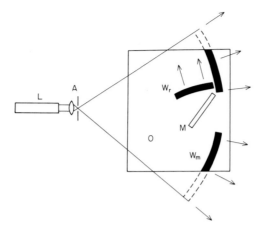

Fig. 7.32. L, laser; A, spatial filter; O, object surface; M, mirror; W_m, main wavefront; and W_r, reflected wavefront.

Figure 7.32 is a drawing of the object surface (O) with an inclined mirror (M) attached and illuminated by the divergent beam from the laser (L). The expected shape of a wavefront (W_m) partly reflected by the mirror (W_r) is indicated.

Figure 7.33(a)–(d) are photographs taken during the reconstruction of a single hologram plate. The total difference in age of the recorded light is of the order of 800 ps. The four photographs were taken with the camera lens close to the plate and with the camera moved from left to right between each exposure so that the time lapse between them corresponds to around 200 ps. In Fig. 7.33(a) the spherical wavefront, several metres to the right of the laser, is moving across the flat, diffuse object surface. The wavefront has just reached the lower left part of the plane mirror attached perpendicularly to the object surface with its normal inclined 40° to the horizontal line. Figure 7.33(b) shows how the wavefront has reached the middle of the mirror and how the light is beginning to be reflected upward and slightly to the left. In Fig. 7.33(c) the wavefront has just passed the mirror and all the reflected light is leaving. In Fig. 7.33(d) the two parts of the wavefront have separated completely, the reflected light leaving a hole in the main wavefront.

The new measuring technique has many advantages over conventional methods of high-speed photography. It produces a three-dimensional, continuous, moving picture of the light in flight. The time resolution is extremely high because no mechanics are involved and there is no electrical capacitance or inductance. If the hologram plate were tilted in relation to the normal to the hyperboloids, each length unit on the plate would represent an even shorter time than that corresponding to the inverse value of the speed

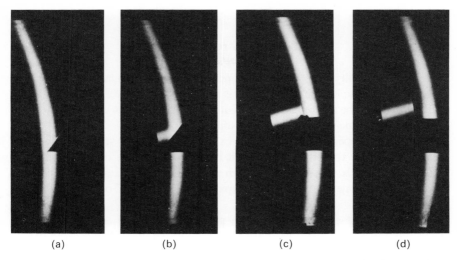

(a)   (b)   (c)   (d)

Fig. 7.33. (a) A spherical wavefront from an argon laser enters at the left, illuminating a white-painted, flat surface at an oblique angle. The lower left-hand end of a tilted mirror is just reached. (b) The wavefront has reached the middle of the mirror, the normal of which is inclined at 40° to the horizontal. The light is reflected upwards and to the left. (c) All the reflected light is separating from the main wavefront, which has just passed the mirror. (d) The two components of the light have separated completely, the reflected light leaving a black hole in the spherical wavefront, which exits to the right.

of light. By forcing the reference beam to travel in a zigzag manner over the plate or along a long hologram film, one could increase the time capacitance of the method by several orders of magnitude. By using the short, picosecond pulse (or pulses) from mode-locked lasers or by using lasers with more than one wavelength, it is possible to increase the resolution to a fraction of a millimetre. By special design of the experiment (for instance, by enlarging the duration of the illumination pulse by the use of one or more scatter plates) not only the light travel but also other ultrahigh-speed events, such as the implosion of deuterium pellets for fusion, could be filmed and even measured interferometrically. For that purpose, what is needed is a laser capable of producing single picosecond pulses. Such a laser was not available at the time of the experiment described above.

It is hoped that this introductory experiment will stimulate interest and open a door to the new possibilities of recording light in flight to study the temporal dynamics of interferometry, refraction, diffraction, and the behaviour of light passing through optical fibres and integrated optics. This and similar new techniques could become powerful tools in the understanding of optics and other ultrahigh-speed phenomena.[12]

7.6 Stability requirements

Two main stability problems can appear in hologram interferometry. The first one is that no holographic image is recorded whatsoever; the second one is that it cannot be trusted that the recorded fringes really are the result of the phenomenon to be studied and of nothing else.

If the plate appears to be correctly exposed but there is still no image, the reason could be that the path length difference between object beam and reference beam is more than the coherence length. This situation can be checked by using a string to compare the distances, or by using the holo-diagram.

If there is still no image after the possible coherence errors have been removed, the reason could be that there is too much unwanted motion of the object during the exposure. In those cases in which it is impossible to make the object more stable, the holographic configuration should be so arranged that the object is positioned in the holo-diagram where the k-value is as high as possible. If the motion of the object is known to be in a certain direction, this direction should be tangential to the ellipsoids of the holo-diagram. If it is still impossible to obtain a hologram, fix the reference mirror to the object as described in Section 3.3.1. The interference fringes of such a hologram can be evaluated like those of an ordinary hologram with one exception: in-plane motions cannot be evaluated by studying fringe motions as the observation is moved along the plate (the dynamic method of Section 4.2.2).

7.6.1 Optical components

Should it still not be possible to make a hologram the reason might be insufficient stability of the optical components (see Figs 3.14 and 3.15). The spatial filter should not be fixed to the laser if this is vibrating, e.g. because of built-in transformers or cooling by fans or streaming water.

When working on a table do not have the power supply on the same table. The shutter should not be supported by the table either. All the components have to be supported in a stable way. A weak support can magnify a small vibration ten or even a hundred times. Thus even a well isolated table might be useless in combination with unstable or elastic mounts of mirrors and other components. If possible, it is advisable to work directly on a basement floor.

Should there still be no hologram recorded, check the temperature gradients and air currents. Often the ventilation of a temperature-controlled room causes rather quick changes between two values. Shut it off during exposure. If working in an ordinary room close all the ventilation points and check for draughts from windows and doors or warm current from heating elements. Sometimes it is necessary to stop air currents by enclosing the light path and the objects using sheets of foam plastic. When working in a large room it can even be necessary to build a tent of black plastic film covering the entire holographic set-up. One advantage is then that exposures can be made without having to darken the whole room.

If, in spite of all these precautions, still no hologram is recorded, a mirror can be fixed to the object at such an angle that this mirror reflects the laser light to the hologram holder. Thus the hologram plate is illuminated by a reference beam and an object beam that are both reflected by mirrors. Remove the hologram holder and set in its place a beam-splitter which is adjusted so that the reflection of the object beam coincides with the reference beam passing through the beam-splitter. Thus, two components of the divergent laser beam are recombined, resulting in circular interference fringes. The motions of these fringes are, of course, identical to the motions of the image-carrying hologram fringes, so that by studying the stability of the circular fringes one can judge the possibility of recording a hologram.

If, finally, even with this interferometer, no fringes are seen, a digital laser interferometer, e.g. of the Hewlet Packard type, can be used to check the stability of each optical component. This interferometer can be connected to a recorder which produces a curve over the vibrations and rigid body motions as a function of time. If the fringes are moving all the time in a random way, the use of a pulsed laser, e.g. a ruby laser, is recommended (Section 7.1.3).

However, let us assume that after some adjustments it has proved possible to make a successful hologram. By the use of a stable hologram holder, such as the one shown in Fig. 3.5, the hologram can, after processing, be repositioned and the real-time fringes studied. As soon as these fringes can be observed it is usually rather easy to improve the holographic set-up because they reveal the instabilities.

It is usually difficult to record a hologram if the real-time fringes move all the time. When they move now and then, but in between return to their original position, the hologram will usually be of reasonably good quality. Even if they move almost all the time around a more or less stable equilibrium, it can be possible to make a hologram because of the averaging process. The necessary condition for making a hologram is that the image-carrying fringes should be at a particular position at least during most of the time of the exposure.

The fringes caused by motions of the object or any optical component are usually straight, while those caused by air currents twist and wringle. It is often found that the latter disturbance has got a period of around 1 s. Thus some of the problems of air turbulence might be avoided either by making a short exposure, e.g. a fraction of a second, or a long exposure, e.g. some 10 s.

7.6.2 Object fixtures

If the stability of the object has to be improved, check that it cannot rock and sway. It is important that it should rest on three points; do not assume that a flat surface is flat, instead use, say, three coins to make sure that there will be just three supports. It is not necessary to clamp the object or the optical components; often it is quite sufficient to let gravity keep them in place. For example, a well clamped mirror might go on vibrating for tens of seconds, while a mirror just leaning against a heavy piece of steel immediately falls back in position after a disturbance. Do not use plasticine or similar materials that can deform slowly under pressure for many hours.

Sometimes the object has to be removed from the holographic set-up in between the two exposures. In that case it is necessary to design a fixture which enables sufficiently accurate repositioning. If real-time or double-exposure holography is used, the position error should usually be no more than a few micrometres while sandwich holography accepts errors of up to around 1 mm.

For high precision it is important that the object is not deformed by clamping forces and that the number of contact points is no higher than that necessary for a defined position. Thus a square box could rest on three contact points along its horizontal bottom side (a, b and c of Fig. 7.34). One side rests against two points (d and e) along a horizontal line while another side rests against one single point (f). In this way the position of the box is defined by six contact points which is a necessary and sufficient number to define the position of any three-dimensional body. (The hologram holder of Fig. 3.5 has also got six contact points.)

The contact points could be steel ball-bearings which are securely fastened to the fixture; they should easily slide against the object surface so that no undefined forces deform the object. The six points are sufficient only if there is a force pressing the object against them. Usually it is enough to let gravity supply this force. The object simply rests on three of the points while it is gently pressed by hand against the other three points during repositioning.

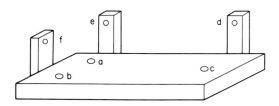

Fig. 7.34. A fixture used for the repositioning of an object in the form of a rectangular box. Its horizontal bottom side rests on three spherical contact points (a, b, c). One side rests against two points (d, e) while a third side rests against one point (f).

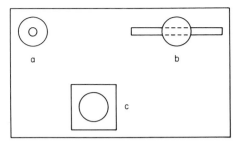

Fig. 7.35. An alternative repositioning fixture. The bottom side of the object has three pins with spherical ends. One rests in a hole (a), the second one rests in a slot of triangular cross-section (b), while the third one rests on a flat surface (c).

Sometimes it is possible to tilt the fixture with its object so that gravity keeps it resting against all the contact points, as in the plate-holder of Fig. 3.5.

There is an alternative configuration for the repositioning of the object. Three pins with spherical ends are fixed on the underside of the object. On the horizontal fixture surface is drilled one shallow hole with a diameter of about 0.7 times that of the pin. This hole (a of Fig. 7.35) defines the position of one of the pins of the object. The next pin should rest in a slot in the form

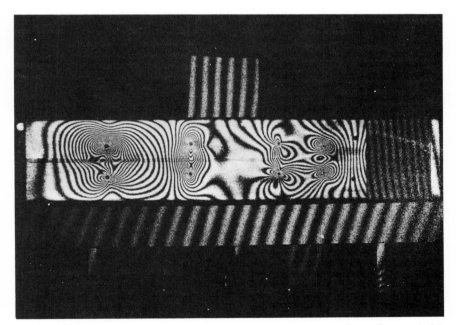

Fig. 7.36. An object, almost 1 m long, was recorded on one plate of a sandwich hologram. It was then transported by car to a workshop where eight holes were drilled. Twenty days later the second exposure was made, resulting in this high quality hologram.[13]

of a trench with a triangular cross-section (b of Fig. 7.35). Thus any changes in dimension of the object or the fixture should result in a sliding action of the pin along the slot, without the introduction of deforming forces. The third pin finally should just rest on a flat surface so that it is free to slide in any horizontal direction. The first pin represents three contact points, the second two and the third one contact point.

Utilizing the repositioning fixture described first, it has proved possible to reposition objects of 1 m in length without difficulties. Figure 7.36 shows the fringes of a sandwich hologram produced by this method. An image was recorded on one plate of the sandwich hologram. The object was then transported by car to a workshop where six holes were drilled. After 20 days it was brought back to the hologram laboratory and the second plate of the sandwich hologram was recorded. The contrast of the fringes is just as good as in an ordinary double-exposure hologram.[13]

This experiment demonstrated that hologram interferometry can be used to study the deformation of machine parts after they have been exposed to forces caused by machining or by their normal use. An airplane structure can be recorded before and after flight in order to find abnormal deformations indicating areas at which there is a risk of future failures.

References

1. N. Abramson. The holo-diagram: a practical device for making and evaluating holograms. *Appl. Opt.* **8**, 1235 (1969).
2. B. Hildebrand and K. Haines. Multiple-wavelength and multiple-source holography applied to contour generation. *J. opt. Soc. Am.* **57**, 155 (1967).
3. T. Tsuruta, N. Shiotake, J. Tsujiuchi and K. Matsuda. Holographic generation of contour map of diffusely reflecting surface by using immersion method. *Jap. J. appl. Phys.* **6**, 66 (1967).
4. N. Abramson. Holographic contouring by translation. *Appl. Opt.* **15**, 1018 (1976).
5. S. Walles. Visibility and localization of fringes in holographic interferometry of diffusely reflecting surfaces. *Ark. Fys.* **40**, 348 (1969).
6. N. Abramson. Sandwich hologram interferometry. 3. Contouring. *Appl. Opt.* **15**, 200 (1976).
7. J. Giordmaine and P. Rentzepis. Two-photon excitation of fluorescence by picosecond light pulses. *Appl. Phys. Lett.* **11**, 216 (1967).
8. M. Dugay. Light photographed in flight. *Am. Scient.* **59**, 551 (1971).
9. D. Attwood, L. Coleman and D. Sweeney. Holographic microinterferometry of laser-produced plasmas with frequency-tripled probe pulses. *Appl. Phys. Lett.* **26**, 616 (1975).
10. N. Abramson. The holo-diagram. III. A practical device for predicting fringe patterns in hologram interferometry, *Appl. Opt.* **9**, 2311 (1970).
11. N. Abramson. Light-in-flight recording by holography. *Opt. Lett.* **3**, 121 (1978).
12. H. Bartelt, S. Case and A. Lohmann. Visualization of light propagation. *Opt. Commun.* **30**, 13 (1979).
13. H. Bjelkhagen. Sandwich holography for compensation of rigid body motion and reposition of large objects. Proceedings of Photo-optical Instrumentation Engineers Conference in California **215**, 85 (1980).

Subject Index